Rapid Assessment Program

Bali Marine Rapid Assessment Program 2011

RAP
Bulletin
of Biological
Assessment
64

Edited by
Putu Liza Kusuma Mustika, I Made Jaya Ratha,
Saleh Purwanto

MARINE AND FISHERIES AFFAIRS
BALI

SOUTH EAST ASIA CENTER
FOR OCEAN RESEARCH AND
MONITORING

WARMADEWA UNIVERSITY

CONSERVATION INTERNATIONAL
INDONESIA

Denpasar, Bali
August 2012

Bali Marine Rapid Assessment Program 2011

First English edition August 2012

Suggested citation:
Mustika, P. L., Ratha, I. M. J. & Purwanto, S. (eds) 2012. T*he 2011 Bali Marine Rapid Assessment (Second English edition August 2012).* RAP Bulletin of Biological Assessment 64. Bali Marine and Fisheries Affairs, South East Asia Center for Ocean Research and Monitoring, Warmadewa University, Conservation International Indonesia, Denpasar. 137 pp.

Photographers:
Gerald R. Allen:
Cover (left side), banner and contents of Chapter 3

Lyndon DeVantier:
Contents of Chapter 5

I Made Jaya Ratha:
Banner of Executive Summary, banner of Chapter 1, banner and contents of Chapter 2, banner of Chapter 6

Mark Erdmann:
Cover (right side), contents of Chapter 3, Figure 5.5

Muh. Erdi Lazuardi:
Banner of Chapter 4, banner of Chapter 5

Emre Turak:
Contents of Chapter 5

Cartographers:
Gerald R. Allen:
Figure 3.19

I Made Jaya Ratha
Figure 1.2

Emre Turak/Lyndon DeVantier:
Figures 5.3, 5.4, 5.14

Nur Hidayat:
Figures 1.1, 5.2, 6.1

Ketut Sudiarta:
Figure 4.1

Translator:
Putu Liza Mustika (Chapters 2 and 4)

Layout:
Kim Meek

Editors:
Putu Liza Mustika
I Made Jaya Ratha
Saleh Purwanto

English editor:
Philippa Blake

Conservation International is a private, non-profit organization exempt from federal income tax under section 501c(3) of the Internal Revenue Code.

ISBN: 978-1-934151-51-8

RAP Bulletin of Biological Assessment was formerly RAP Working Papers. Numbers 1–13 of this series were published under the previous series title.

FOREWORD FROM THE GOVERNOR OF BALI

With the blessing of The Almighty, we hereby release the Bali Marine Rapid Assessment Program report. The report contains the current status of all of Bali's coastal and marine resources, and becomes a key reference as the Bali Provincial Government plans for the management of our coast and sea, as mandated by the Local Regulation 16/2009 on the 2009-2029 Bali Spatial Plan.

We take pride in these results that show that Bali's marine resources remain exceptional. The report presents comprehensive analyses of the island's coral reef and reef fish. Currently recovering from climate change impacts, Bali's coral reef ecosystem is highly diverse. Similarly, the reef fish communities in Bali are also highly diverse. Bali's underwater beauty remains enigmatic; indeed this process has identified several new species. Bali's coral reefs and other reef ecosystem organisms are important underwater economic resources, for fisheries, food security, and marine tourism. The discovery of several new fish species and one new coral species reinforce the truly remarkable nature of Bali's underwater treasures.

We must also pay particular attention to some serious matters, however, including the declining populations of economically important fish species, and conflict over resource use. We must consult Law 27/2007 on the management of Coastal Zones and Small Islands and the Local Regulation 16/2009 on the Bali Spatial Plan to control these problems and to minimise the emergence of new problems. These regulations can guide the development of the Strategic Plan for the Management of Bali Coastal Zones, a Management Plan and an Action Plan for Coastal Zones and Small Islands to foster sustainable fisheries and marine tourism in Bali. These efforts must be paired with strong commitment and collaboration from partners in Government, the private sector and local communities in management actions such as surveillance and law enforcement.

I hereby extend my gratitude to all who have actively been involved in the marine rapid assessment. This report should be used for our marine and fisheries resources. I hope all government agencies, private sector partners and other stakeholders can use this information to guide development which is integrated with nature.

I hope that this report inspires all stakeholders to build a better Bali for the future - an advanced, safe, peaceful and prosperous Bali.

With thanks,

Om Shanti, Shanti, Shanti Om.

THE GOVERNOR OF BALI

MADE MANGKU PASTIKA

FOREWORD FROM THE GOVERNOR OF BALI (INDONESIAN VERSION)

GUBERNUR BALI

KATA PENGANTAR

Om Swastyastu,

Puja pangastuti angayubagya kita panjatkan kehadapan Ida *Sang Hyang Widhi Wasa* / Tuhan yang Maha Esa, atas *asung kertha wara nugraha*-Nya Laporan "Bali Marine Rapid Assesment Program (Bali MRAP)" ini dapat diterbitkan. Laporan sebagai hasil kajian potensi pesisir dan kelautan ini merupakan salah satu referensi Pemerintah Provinsi Bali dalam merencanakan pengelolaan wilayah pesisir dan laut, sesuai amanat Peraturan Daerah Provinsi Bali Nomor 16 Tahun 2009 tentang Rencana Tata Ruang Wilayah Provinsi Bali.

Kita patut berbangga, laporan ini menunjukkan bahwa Bali memiliki potensi kelautan yang sangat tinggi. Secara komprehensif disajikan kondisi terumbu karang dan ikan karang di Bali. Ekosistem terumbu karang di Bali sangat bervariasi, dan sedang mengalami proses pemulihan yang pesat dari dampak dan perubahan iklim global. Ikan karang di Bali sangat beranekaragam bahkan dengan indeks densitas yang tinggi. Keindahan alam bawah laut Bali memiliki "misteri" yang mempesona dan ditemukan adanya beberapa spesies baru. Potensi terumbu karang di Bali dan biota laut yang hidup di dalamnya merupakan sumber ekonomi bawah laut, baik pengembangan usaha perikanan untuk keamanan pangan (*food security*) bernilai tinggi maupun pariwisata bahari yang mumpuni.Demikian pula penemuan beberapa spesies ikan baru dan satu spesies karang baru di perairan Nusa Penida membuka cakrawala pandang kita bahwa kekayaan laut kita luar biasa.

Kita pun harus turut mencermati beberapa permasalahan serius yang dijumpai di lapangan, antara lain: polusi, menurunnya spesies ikan yang memiliki nilai ekonomis, serta munculnya konflik kepentingan dalam pemanfaatan kawasan. Sementara untuk mencegah meluasnya permasalahan atau mungkin munculnya permasalahan baru, sesuai amanat Undang-Undang Nomor 27 Tahun 2007 Tentang Pengelolaan Wilayah Pesisir dan Pulau-pulau Kecil serta arahan Peraturan Daerah Provinsi Bali Nomor 16 Tahun 2009 Tentang Rencana Tata Ruang Wilayah Provinsi Bali, perlu dipertimbangkan penyusunan Rencana Strategis Pengelolaan Wilayah Pesisir Bali, Rencana Zonasi, Rencana Pengelolaan, dan Rencana Aksi Wilayah Pesisir dan Pulau-pulau Kecil Bali untuk mendorong pembangunan bidang perikanan dan juga bidang pariwisata secara berkelanjutan di Bali. Upaya tersebut harus diikuti komitmen dan kerjasama yang baik antara Pemerintah, sektor swasta dan masyarakat local, termasuk dalam melakukan pengawasan dan penegakan hukum.

Saya menyampaikan terima kasih kepada semua pihak yang telah terlibat aktif dalam melaksanakan kajian kelautan ini. Selanjutnya laporan ini bisa dimanfaatkan sebagai database sumberdaya kelautan dan perikanan yang kita miliki. Saya berharap seluruh instansi pemerintah maupun swasta serta pemangku kepentingan lainnya mampu memanfaatkan data dan informasi ini dalam menyusun program pembangunan secara terintegrasi.

Sebagai akhir kata, semoga laporan ini memberikan inspirasi bagi semua pihak untuk menjadikan Bali ke depan lebih baik, sebagaimana cita-cita mewujudkan masyarakat Bali yang maju, aman, damai dan sejahtera.

Sekian dan terima kasih.

Om Shanti, Shanti, Shanti Om.

GUBERNUR BALI,

MADE MANGKU PASTIKA

FOREWORD FROM THE HEAD OF MARINE AND FISHERIES AFFAIRS BALI

BALI PROVINCE GOVERNMENT
MARINE AND FISHERIES SERVICE AFFAIRS
Patimura Sreet No. 77 Telp. (0361) 227926 Fax. (0361) 223562
DENPASAR

FOREWORD FROM THE HEAD OF MARINE AND FISHERIES AFFAIRS BALI

Environmental and biodiversity conservation is an important consideration in achieving Bali's vision as a Clean and Green Province. Better understanding the richness of our natural resources is imperative for designing the island's sustainable development.

The Bali Marine Rapid Assessment Program (Bali MRAP) has been conducted by the Provincial Government of Bali through its Marine and Fisheries Agency. Funded by the USAID in 2011, the assessment program was coll aboration among several partners. In a relatively short period, the assessment produced relevant data and information on Bali's marine richness, particularly from the perspectives of coral reefs and reef fishes.

The assessment did not cover the entire coastal waters of Bali. However, the results still provide valuable input for the government to manage Bali's marine species and to plan the development of Marine Protected Areas at the regency level. We also hope that other related stakeholders on the island will use the report to plan and manage their resources in a sustainable manner.

The first edition of this report was issued in November 2011. This book, the second edition of the report, contains some updates including the published names of new reef fish species. We extend our gratitude to all who have supported the Bali MRAP, including its publication processes.

Denpasar, 1st August 2012

PEMERINTAH PROPINSI BALI
DINAS KELAUTAN DAN PERIKANAN

Ir. I MADE GUNAJA, M.Si.
FIRST LEVEL SUPERVISOR
NIP. 19640620 199003 1 012

Foreword

Executive Director of Conservation International Indonesia

In addition to being famous for its nature-based tradition and culture, Bali is also a renowned world-class tourist site. Located at the southwestern corner of the Coral Triangle, the waters of Bali are home to abundant marine life that supplies provisions for the local community and that has become a major tourism attraction.

However, managing Bali's coastal and natural resources is a big challenge. Rapid coastal development has not yet been balanced by a proper long-term management plan. It is plausible, therefore, to question Bali's long-term economic sustainability from this perspective.

The government, private sector and NGOs of Bali have all initiated various strategies for the island's long-term development. The hard work and collaborative efforts of these parties resulted in the issuance of Local Regulation 16/2009 (on Bali spatial plan), which has become the guide for planning in Bali over the next 20 years. The establishment of Marine Protected Areas (MPAs) and an MPA network in Bali is but one strategy to translate the aforementioned Bali spatial plan.

Accordingly, through the Bali MPA Network program, Conservation International Indonesia (CII) aims to have the island's marine and coastal resources managed effectively to protect ecosystem and socio-economic services for the local community and government. By aiming to develop an effectively managed MPA network for Bali that supports the government's policies of 'one island, one management' and 'Bali Clean and Green Province', CII hopes to facilitate collaboration between the government and all related stakeholders. CII also hopes that, supported by capable and professional MPA managers, the Bali MPA Network will serve as the main support for resilient marine tourism on the island.

During a stakeholder workshop for the development of a Bali MPA network in June 2010, 25 locations were nominated as potential MPA sites. Scientific input (bio-ecological and socio-economic) was deemed crucial to complement the MPA network design, prompting the Bali government and CII to conduct a comprehensive assessment of the island's marine environment led by Bali's Marine and Fisheries Agency. We hope that the results of the 2011 Marine Rapid Assessment Program will serve as a scientific resource to guide the future Bali MPA Network team to work towards a Green Economy and Sustainable Tourism.

We thank the Bali government (the Marine and Fisheries Agency in particular) for its support and leadership in this assessment. We also thank the members of the Bali MRAP team: the Indonesian Ocean Institute, the Marine and Fisheries Agency, Warmadewa University, Udayana University, the Bali Nature and Conservation Agency (BKSDA), the Bali Diving Academy and others. It is our hope that this assessment and the following recommendations are useful for the decision makers and marine and coastal managers of Bali.

Denpasar, 24 October 2011

Ketut Sarjana Putra
Country Executive Director
Conservation International Indonesia

Table of Contents

List of Figures

List of Plates

List of Plates, *continued*

List of Tables

Participants

I Gusti Putu Ngurah Nuriartha
(Senior Responsible Officer and Advisor)
Marine and Fisheries Affairs of the Bali Province
Jl. Patimura 77 Denpasar Bali
Fax. (0361) 223562

Eghbert Elvan Ampou
(Expert on coral reef ecology)
South East Asia Center for Ocean Research and Monitoring Bali
Jl. Baru Perancak-Jembrana, Bali
Fax. 0365-44278
Email: elvan_ampou76@yahoo.com

Muhammad Erdi Lazuardi
(Expert on coral reef ecology)
Conservation International (CI) Indonesia
Jl. Dr Muwardi 17 Renon, Denpasar Bali 80235
Fax. +62 361 235 430
Email: m.lazuardi@conservation.org

Gerald Robert Allen
(Reef fish expert)
Conservation International
1919 M Street NW, Suite 600
Washington, DC 20036, USA

Emre Turak
(Coral expert)
Conservation International
1919 M Street NW, Suite 600
Washington, DC 20036, USA

I Made Jaya Ratha
(Coastal socio-economic expert)
Conservation Indonesia (CI) Indonesia
Jl. Dr Muwardi 17 Renon, Denpasar Bali 80235
Fax. +62 361 235 430
Email: i.ratha@conservation.org

Ketut Sarjana Putra
(Senior Responsible Officer)
Conservation International (CI) Indonesia
Jl. Dr Muwardi 17 Renon, Denpasar Bali
80235
Fax. +62 361 235 430
Email: k.putra@conservation.org

Mark Van Nydeck Erdmann
(Field Coordinator and reef fish expert)
Conservation International (CI) Indonesia
Jl. Dr Muwardi 17 Renon, Denpasar Bali
80235
Fax. +62 361 235 430
Email: mverdmann@gmail.com

Suciadi Catur Nugroho
(Expert on coral reef ecology)
South East Asia Center for Ocean Research
and Monitoring Bali
Jl. Baru Perancak-Jembrana, Bali
Fax. 0365-44278
Email: suciadi_cn@yahoo.com

I Ketut Sudiarta
(Expert on coral reef ecology)
Warmadewa University
Jl. Akasia 10 Denpasar, Bali
Email: ikt_sudiarta@yahoo.co.id

Lyndon DeVantier
(Coral expert)
Conservation International
1919 M Street NW, Suite 600
Washington, DC 20036, USA

INTRODUCTION

The Indonesian province of Bali is located just east of the island of Java and comprises 563,666 ha covering the main island of Bali and the smaller satellite islands of Nusa Penida, Nusa Lembongan, Nusa Ceningan, Pulau Serangan and Pulau Menjangan. Bali is known throughout the world for its unique Hindu culture and as a top global tourism destination. It is also situated in the southwest corner of the Coral Triangle—the region of the highest marine biodiversity on the planet. Bali's rich marine resources have long been an important economic asset to the island—both as a source of food security for local communities (many of whom derive a significant proportion of their animal protein needs from seafood) and also as a focus for marine tourism. Diving and snorkeling attractions such as Nusa Penida, Candi Dasa, Menjangan Island (Bali Barat National Park), and the Tulamben *USS Liberty* wreck have been drawing tourists into Bali's waters for decades, while more recently the private marine tourism sector has expanded the menu of options to include sites like Puri Jati, Karang Anyar, and Amed. Other important economic activities in Bali's coastal zone include seaweed farming and ornamental fish collecting.

Unfortunately, despite Bali Governor's Decree No. 324/2000 mandating the implementation of integrated coastal management in the province, rapid and largely uncoordinated development in Bali's watersheds and coastal areas, along with a lack of clear marine spatial planning for the island, has led to significant deterioration of many marine environments around Bali. This is due to a combination of overfishing and destructive fishing, sedimentation and eutrophication from coastal development, sewage and garbage disposal at sea, and dredging/reef channel development. At this point in time, the long-term sustainability of the many important economic activities occurring in Bali's coastal zone is in question.

The Bali provincial government has realized these threats and is now working hard to develop a comprehensive long-term development strategy for the island, including greatly improving spatial planning in both the terrestrial and marine areas of Bali. One important part of this initiative has been the decision by the Bali provincial government to design and implement a comprehensive and representative network of Marine Protected Areas (MPAs) around the island that prioritizes sustainable and compatible economic activities (including marine tourism, aquaculture and sustainable small-scale fisheries).

To initiate the planning for this network of MPAs, the government held a multistakeholder workshop in June 2010. The workshop was organized by the Marine Affairs and Fisheries Agency of Bali Province, in collaboration with the Bali Natural Resources Conservation Agency (KSDA), Warmadewa University, Udayana University, United States Agency for International Development (USAID), Conservation International (CI) Indonesia and local NGOs within the framework of a "Bali sea partnership". The Bali MPA Network workshop was attended by 70 participants from the provincial government, regency governments, universities, NGOs, private sector, community groups, traditional village groups and fishermen groups.

Importantly, the workshop participants identified 25 priority sites around Bali as the top candidates for inclusion in a network of MPAs for the island. This list of sites included existing national/local protected areas such as Bali Barat National Park/Menjangan Island, Nusa Penida, and Tulamben, while also including a number of additional sites that currently have no formal protection. Later, the 25 priority sites were short-listed into seven MPA candidate sites (see Chapter 6 for the list).

In order to move this MPA network agenda forward, the Bali government (in particular, the provincial Marine Affairs and Fisheries Agency) in early 2011 requested the assistance of Conservation International Indonesia's marine program in leading a small team of local and international experts to survey the candidate MPA sites identified by the June 2010 workshop and provide clear recommendations on priority development sites and next steps for the design of the MPA network. The team was asked to build upon the survey data compiled during the November 2008 CI-led "Marine Rapid Assessment" of the Nusa Penida reef system to provide a more comprehensive report on the biodiversity, community structure, and current condition of coral reefs and related ecosystems around Bali. Based upon this information, the team was also requested to provide recommendations on how to best prioritize the 25 candidate sites for inclusion in an ecologically-representative network of MPAs.

THE OBJECTIVES OF THE BALI MARINE RAPID ASSESSMENT PROGRAM

The assessment, conducted from 29 April–11 May 2011, had the following three primary objectives:

- Assess the current status (including biodiversity, coral reef condition and conservation status/resilience of hard corals and coral reef fishes) of the majority of the 25 candidate MPA sites around the island of Bali identified by the June 2010 Bali MPA Network workshop, compiling thorough species-level inventories for each site.
- Compile spatially-detailed data on biological features which must be taken into consideration in finalizing the Bali MPA Network design. This design includes not only an analysis of any differences in reef community structure of the priority sites, but also specifically identifying areas of outstanding conservation importance due to rare or endemic hard coral or fish assemblages, presence of reef fish spawning aggregation or cleaning sites, reef communities exposed to frequent cold-water upwelling that are resilient to global climate change, or other outstanding biological features.
- Taking the above into account, provide concrete recommendations to the Bali government on the next steps to be taken to finalize the design of the Bali MPA Network.

SURVEY RESULTS: GENERAL

- The MRAP was successfully completed during a 13-day period of 29 April–11 May 2011, with a presentation of preliminary findings to the Governor of Bali on 12 May 2011. The survey team comprised 12 individuals, including representatives of the Marine and Fisheries Affairs of the Bali Province (DKP), the Perancak Research Centre for Oceanography (BROK), Warmadewa University, and six local and international marine taxonomists from Conservation International. The survey was funded in its entirety through the Coral Triangle Support Program (CTSP) of the USAID.
- In total, 33 sites were successfully surveyed (see Table 1), representing the majority of the 25 candidate MPA sites that were previously identified by the June 2010 expert workshop. The survey began at the southern tip of Bali and proceeded in a counter-clockwise fashion around the island until the northwest corner was reached—at which point the survey team could not continue any further down the west coast due to wave conditions making it impossible to dive. The data from these 33 sites have been combined with previous data taken from 19 sites during the November 2008 Nusa Penida MRAP; coral and reef fish taxonomic and community structure analysis presented in the report thereby utilize this combined data set for 52 sites.
- With over 350 man-hours of diving conducted during the MRAP, the team was overall very impressed with both the surprisingly high biodiversity (including a number of new species) and especially the finding that Bali's coral reefs are in a very active stage of recovery after the coral bleaching, destructive fishing and crown-of-thorns starfish outbreaks that were largely thought to have decimated these reefs in the late 1990s and into 2001. The impressive ratio of live to dead hard coral recorded of 7:1 is extremely encouraging and is testament to the resilience of these reefs. At the same time, the team found abundant evidence of serious management problems for Bali's marine ecosystems, such as: omnipresent plastic trash; severe signs of overfishing including near complete depletion of reef sharks and large commercially valuable fishes such as Napoleon wrasse; and serious resource use conflicts between communities engaged in marine tourism livelihoods and outside fishing interests unsustainably targeting the very resources upon which marine tourism depends.

SURVEY RESULTS: REEF FISH BIODIVERSITY

- The biodiversity of reef fishes was assessed for 29 of the 33 survey sites by G. Allen and M. Erdmann, using underwater visual census from a 1–70m depth. A total of 805 species were recorded during the survey. When combined with results of the previous Nusa Penida MRAP, the total diversity recorded for the Bali region includes 977 species of reef fish representing 320 genera and 88 families.
- Wrasses (Labridae), damselfishes (Pomacentridae), gobies (Gobiidae), cardinalfishes (Apogonidae), groupers (Serranidae), butterflyfishes (Chaetodontidae), and surgeonfishes (Acanthuridae) are the most speciose families on Bali reefs with 114, 96, 84, 59, 54, and 39 species respectively.
- Species counts at individual sites ranged from 42 to 248, with an average of 153 species per site. Those sites with the highest diversity recorded included Anchor Wreck, Menjangan (248 species), Batu Kelibit, Tulamben (246 species), Kepah, Amed (230 species), Jemeluk, Amed (220 species) and Bunutan, Amed (217 species).

- The majority of Bali fishes have broad distributions in the Indo-Pacific (56.4%) or western Pacific (25.3%). Minority categories of special interest involve species that are mainly distributed in the Indian Ocean (3%) and Indonesian endemics (3.3%). A total of 16 reef fish species are now currently only known from Bali and the nearby Nusa Tenggara Islands to the east and are considered local endemic species.
- At least 13 new and undescribed reef fish species were recorded and collected during the survey, including two fang blennies (*Meiacanthus*), two jawfish (*Opistognathus*), three dottybacks (*Pseudochromis* and *Manonichthys*), a dartfish (*Ptereleotris*), a clingfish (*Lepidichthys*), a grubfish (*Parapercis*), a cardinalfish (*Siphamia*), and two gobiids (*Grallenia* and *Priolepis*). Though most of these undescribed species have been previously recorded from surrounding regions, five were recorded for the first time during the two MRAP surveys.

Table 1. Summary of survey sites for Bali MRAP 29 April – 11 May 2011.

Site No.	Date Surveyed	Location Name	Coordinates
1	29 April 11	Terora, Sanur (Grand Mirage)	08° 46.228' S, 115° 13.805' E
2	29 April 11	Glady Willis, Nusa Dua (Grand Mirage)	08° 41.057' S, 115° 16.095' E
3	29 April 11	Sanur Channel	08° 42.625' S, 115° 16.282' E
4	30 April 11	Kutuh Temple, Bukit	08° 50.617' S, 115° 12.336' E
5	30 April 11	Nusa Dua	08° 48.025' S, 115° 14.356' E
6	30 April 11	Melia Bali, Nusa Dua	08° 47.608' S, 115° 14.192' E
7	1 May 11	West Batu Tiga (Gili Mimpang)	08° 31.527' S, 115° 34.519' E
8	1 May 11	East Batu Tiga	08° 31.633' S, 115° 34.585' E
9	1 May 11	Jepun (Padang Bai)	08° 31.138' S, 115° 30.619' E
10	2 May 11	Tepekong (Candidasa)	08° 31.885' S, 115° 35.167' E
11	2 May 11	Gili Biaha/Tanjung Pasir Putih	08° 30.270' S, 115° 36.771' E
12	3 May 11	Seraya	08° 26.010' S, 115° 41.274' E
13	3 May 11	Gili Selang North	08° 23.841' S, 115° 42.647' E
14	3 May 11	Gili Selang South	08° 24.079' S, 115° 42.679' E
15	4 May 11	Bunutan, Amed	08° 20.731' S, 115° 40.826' E
16	4 May 11	Jemeluk, Amed	08° 20.221' S, 115° 39.617' E
17	4 May 11	Kepa, Amed	08° 20.024' S, 115° 39.244' E
18	5 May 11	Batu Kelibit, Tulamben	08° 16.696' S, 115° 35.826' E
19	5 May 11	Tukad Abu, Tulamben	08° 17.603' S, 115° 36.599' E
20	6 May 11	Gretek, Buleleng	08° 08.969' S, 115° 24.733' E
21	6 May 11	Penutukang, Buleleng	08° 08.270' S, 115° 23.622' E
22	7 May 11	Puri Jati, Lovina	08° 11.032' S, 114° 54.869' E
23	7 May 11	Kalang Anyar, Lovina	08° 11.344' S, 114° 53.841' E
24	8 May 11	Taka Pemuteran	08° 07.775' S, 114° 40.007' E
25	8 May 11	Sumber Kima	08° 06.711' S, 114° 36.451' E
26	9 May 11	Anchor Wreck, Menjangan	08° 05.467' S, 114° 30.131' E
27	9 May 11	Coral Garden, Menjangan (transect only)	08° 05.485' S, 114° 30.486' E
28	9 May 11	Post 2, Menjangan	08° 05.813' S, 114° 31.608' E
29	10 May 11	Secret Bay, Gilimanuk	08° 10.862' S, 114° 26.544' E
30	10 May 11	Secret Bay Reef North, Gilimanuk	08° 09.771' S, 114° 27.116' E
31	11 May 11	Klatakan Pearl Farm 1	08° 13.911' S, 114° 27.249' E
32	11 May 11	Klatakan Pearl Farm 2	08° 14.000' S, 114° 27.463' E
33	8 May 11	Pura Pulaki (reef fish survey only)	08° 08.719' S, 114° 40.756' E

- Though Bali hosts an astounding diversity of fishes for its size, we also found strong signs of overfishing at nearly every site, with large reef fishes of commercial value nearly absent. In over 350 man-hours of diving, the survey team only recorded a grand total of three reef sharks (only at Gili Selang and Menjangan), three Napoleon wrasse (*Cheilinus undulatus*; observed only at Gili Selang and Tulamben), and four coral trout of the genus *Plectropomus*. Equally concerning, the team only recorded a grand total of five marine turtles observed during the survey.

- From the perspective of reef fish assemblage structure, Bali is broadly divisible into four major zones: Nusa Penida, east coast (facing Lombok Strait), north coast, and Secret Bay (Gilimanuk). The developing Bali MPA Network should strive to include representative sites from each of these zones. Other than those areas already included in MPAs(including Menjangan, Nusa Penida, Tulamben and Amed), survey sites worthy of specific conservation focus (based on remarkable fish diversity and excellent habitat conditions) include Batu Tiga, Gili Selang, Taka Pemuteran, Sumber Kima, and Secret Bay (Gilimanuk).

SURVEY RESULTS: HARD CORAL BIODIVERSITY

- Combining the 2008 Nusa Penida and 2011 Bali mainland MRAPs, a total of 85 sites (adjacent deep and shallow areas) at 48 stations (individual GPS locations) were surveyed for hard coral biodiversity. Coral communities were assessed in a broad range of wave exposure, current and sea temperature regimes, and included all main habitat types: cool water rocky shores, cool water reefs with broad flats, warm water reefs with broad to narrow flats, and coral communities developed on predominantly soft substrate.

- Bali hosts a diverse reef coral fauna, with a confirmed total of 406 reef-building (hermatypic) coral species. An additional 13 species were unconfirmed, requiring further taxonomic study. At least one species, *Euphyllia* sp. nov. is new to science, and a second, *Isopora* sp., shows significant morphological difference from described species, such that there are likely to be more than 420 hermatypic Scleractinia present, in total. To place this in perspective with other regions surveyed within the Coral Triangle, this figure for overall coral species richness is similar to those from Bunaken National Park and Wakatobi (392 and 396 spp. respectively), significantly higher than for Komodo and Banda Islands (342 and 301 spp. respectively), and lower than Derawan, Raja Ampat, Teluk Cenderwasih, Fak-Fak/Kaimana and Halmahera (all with ca. 450 spp. or more).

- Within-station (point) richness around Bali averaged 112 species (s.d. 42 spp.), ranging from a low of just two species (at Station B22, a muddy non-reefal location) to a high of 181 species at B16 (Jemeluk, Amed). Other species-rich stations included the Anchor Wreck, Menjangan (168 spp., Station B26) and Penutukang (164 spp., Station B21).

- Using cluster analysis at the station level, five major coral community types were identified, related to levels of exposure to waves, currents/upwelling, substrate type and geographic location. They include: the relatively sheltered north coast of Bali (Menjangan to Amed); wave exposed reefs along south Bali, south Nusa Penida and northwest Bali; clear-water, current-swept reefs of northern Nusa Penida (and several reefs of east Bali); fringing reefs of east Bali from Nusa Dua to Gili Selang; and several predominantly soft-bottom, marginal reef habitats including Puri Jati, Kalang Anyar and Secret Bay (Gilimanuk). These five communities were further sub-divided into ten main coral assemblages, with each of the five communities characterized by a more-or-less distinctive suite of species and benthic attributes.

- Cover of living hard corals averaged 28%. Dead coral cover was typically low, averaging $<4\%$ overall, such that the overall ratio of live to dead cover of hard corals was highly positive (7:1), indicative of a reef tract in moderate to good condition in terms of coral cover. Areas of high soft coral cover occurred on rubble beds, likely created by earlier destructive fishing, coral predation and the localized dumping of coral down-slope during creation of algal farms. Minor evidence of recent and not-so-recent blast fishing and coral diseases was also present, the latter typically on tabular species of *Acropora*. Some localized damage from recreational diver impacts was also apparent. A very strong stress response (in the form of cyanobacterial growth) of corals in the southeast of Bali (Sanur and Nusa Dua) was likely linked with eutrophication and sewage seepage from coastal tourism development.

- Bali's coral faunal composition is typical of the larger region, with most species recorded being found elsewhere in the Coral Triangle. The overall high similarity in species composition with other parts of Indonesia notwithstanding, several important differences were apparent among these regions in the structure of their coral communities. Bali showed closest similarity to Komodo, also in the Lesser Sunda Islands and subject to a somewhat similar environmental regime in respect of current flow and cool water upwelling. These regions showed moderate to high levels of dissimilarity from most other regions to the north, notably from the more species- and habitat-rich regions of Derewan, Sangihe-Talaud, Halmahera and the Bird's Head Seascape of West Papua.

- Discovery of an undescribed species of *Euphyllia* on the east coast of Bali, and the presence of other apparently local endemic corals, notably *Acropora suharsonoi*, suggests that the region does have a degree of faunal uniqueness, possibly related to the strong current flow through Lombok Strait. Given this situation, a precautionary approach demands that Bali's reefs require careful management of local impacts, as replenishment from outside sources may be a prolonged process.

- Reefs of particularly high conservation value around mainland Bali were widespread along the east and north coasts, and include Jemeluk, Menjangan, Gili Tepekong, Penutukang, Bunutan, Gili Selang and Gili Mimpang. Coral communities of Nusa Penida differ significantly from those of the mainland of Bali, and are subject to different environmental conditions and human uses, and hence may require separate management focus. Reefs of high local conservation value around Nusa Penida include those at Crystal Bay, Toya Pakeh, Sekolah Dasar and Nusa Lembongan.

- The wave-exposed south coast was not thoroughly surveyed because of large ocean swell. Many of the south coast reefs are highly prized for surfing, and as such draw large numbers of tourists to Bali each year. In the latter respect, their future conservation should be considered a priority for maintaining surf tourism on the island (noting that the surf conditions are generated by the shallow reefs of the area). Further offshore, the south coast also holds crucial migration corridors for cetaceans and other species.

- The presence of cool water upwelling and/or strong consistent current flow in some areas (especially Nusa Penida and eastern Bali) may be particularly important in buffering the incident reefs against rising sea temperatures associated with global climate change. As such, the Bali MPA Network should strive to include a significant percentage of these reefs within it to best ensure climate change resilience in the network design.

SURVEY RESULTS: CORAL REEF CONDITION

- Coral reef condition was assessed at 27 of the survey sites utilizing a modified "point-intercept transect" method. Two 50-meter transects were placed longitudinally along the reef face at each of two depths (5–7m and 10–14m) for a total of four transects per site. The reef benthos was recorded at 50cm intervals along each transect, using categories of live hard coral (identified to genus level), soft coral, algae, other living benthos (eg, sponges, zoanthids), dead standing coral, coral rubble, and abiotic substrate (eg, sand, rock, silt). Percentage cover of each of these substrate categories was then calculated, along with an index of coral mortality that compares the percentage of living and dead hard coral.

- At the 5–7m depth range, percentage of living hard coral ranged from 21.5–68%, averaging 45.3%. The highest percentage live coral at this depth was found at the Anchor Wreck (Menjangan), with the lowest at East Klatakan. At this depth range, live hard coral was the predominant substrate cover, followed by an average of 17.3% "abiotic substrate" and 11.3% coral rubble.

- At the 10–14m depth range, percentage of living hard coral ranged from 11–76%, with the highest cover recorded at Gili Tepekong and the lowest at Kutuh. On average, reefs at this depth range were dominated by hard coral (32.8%), followed by abiotic substrate (21.7%), soft coral (14.9%), and rubble (13.6%).

- In combining the results of both depth ranges, Bali's reefs have an average live hard coral cover of 38.2%. Overall averages of other substrate types in descending order of abundance include: abiotic substrate (20.6%), rubble (12.6%), soft coral (12.1%), other living fauna (6.8%), algae (5.2%), and dead standing coral (4.6%).

- A total of 54 hard coral genera were recorded in the transect surveys, with three genera overall dominating the reefs of Bali: *Acropora* (averaging 9.67% total coverage on each reef), *Porites* (8.12%) and *Montipora* (3.92%).

- If hard and soft coral coverage is combined to give percentage live (hard + soft) coral cover, reefs on mainland Bali showed a range of 31.5–85% (average 54.2%) live coral cover in the 5–7m transects, with the highest figure recorded for Coral Garden, Menjangan and the lowest at Sumber Kima. In the 10–14m depth transects, live coral cover ranged from 12–80.5% (average 47.7%), with the highest recorded at Nusa Dua and lowest at Tukad Abu. It is important to note that, while soft coral is pleasant for divers to look at and does provide shelter and food to certain reef organisms, it does not lay down permanent skeleton (ie, it does not contribute to reef-building) and hence a high percentage of soft coral cover is not preferable for the long-term maintenance of reef structures.

- An index of coral mortality (0 meaning 100% live and 1 meaning 100% dead coral) was calculated for each reef and ranged between .02 and .56 for Bali mainland reefs, with an average of .24, further confirming that Bali's reefs are currently in a state of active recovery from past major mortality events caused by coral bleaching and crown-of-thorns starfish outbreaks.

SURVEY RECOMMENDATIONS:

- Based upon the results of this survey, it is clear that the province of Bali should take decisive action to set up a network of multiple-use MPAs designed to protect the long-term sustainability of local community fisheries livelihoods as well as the burgeoning marine tourism industry. MPAs must be designed, gazetted and managed with the strong participation of local coastal communities, tourism operators and civil society groups, and should be explicitly embedded within a coastal/marine spatial planning framework that seeks to minimize resource user conflict while clearly prioritizing those economic activities that are the most sustainable and benefit the greatest percentage of Bali's communities.

- In order to ensure the recovery of larger reef fishes that both provide an important protein source for local communities while also serving as a primary attraction for divers and snorkelers, these MPAs must include significant "no-take" areas where all forms of fishing and resource extraction are prohibited in order to allow a refuge for these stocks to recover, grow and reproduce—thereby repopulating the reefs of Bali and eventually providing increased catches to fishers operating outside of these no-take areas. To be effective, these no-take areas need to cover 20–30% of the important marine habitats of Bali.

- In designing the Bali MPA Network, it is very important to ensure that all major fish and coral community types are represented in the network—both to protect the full range of Bali's marine biodiversity while also providing the greatest insurance for climate change resilience/adaptation. The results of the Bali MRAP suggest there are at least five major coral community types (fish assemblages largely follow this pattern as well) around Bali, roughly dividing as: northern Nusa Penida; east coast of Bali from Nusa Dua to Gili Selang; north coast reefs of Bali from Amed to Menjangan; north coast soft-bottom habitats in Puri Jati/Kalang Anyar and Gilimanuk Secret Bay; and high wave energy environments on the west and south coasts of Bali and south coast of Nusa Penida.

- In addition to capturing the representativeness of these five major community types, the Bali MPA Network should also strive to include sites of specific high conservation value due to extraordinary biodiversity, particularly intact habitats, rare or endemic species, or fish spawning aggregations or aggregations for cleaning or nesting. Specific sites recognized by the MRAP team as being of particularly high conservation value on the Bali mainland include Batu Tiga (Gili Mimpang), Tepekong, Gili Selang, Tulamben, Amed (Jemeluk and Bunutan), Menjangan, Penutukang, Taka Pemuteran, Sumber Kima, and Secret Bay (Gilimanuk). Previous sites identified of high conservation value in the Nusa Penida MRAP (both due to high biodiversity and for importance as cleaning sites for oceanic sunfish and manta rays) include Crystal Bay, Toya Pakeh, Manta Point, North Lembongan, Batu Abah and Sekolah Dasar (Penida). The turtle nesting beach of Perancak was also previously identified as being of particularly high conservation value. Each of these sites should at least be included in MPAs within the Bali network, and should in fact be strongly considered for inclusion in no-take zones where all fishing and resource extraction is prohibited.

- As a final criterion in site selection/prioritization for the Bali MPA Network, reefs on the east coast of Bali (particularly around Candidasa and Padang Bai) and Nusa Penida are considered particularly important to include within the MPA network from the perspective of climate change resilience; these reefs are subject to both strong currents as well as frequent cold-water upwelling (both a result of the oceanographic feature of the Indonesian Throughflow moving through the Lombok Strait) that should help minimize the effects of warming from climate change.

- Based upon survey findings of only three reef sharks observed in 350 hours of diving, as well as recent evidence of the wholesale slaughter of pregnant female thresher sharks in the waters between Padang Bai and Nusa Dua, the Bali government should strongly consider implementing legislation to create a shark sanctuary in Bali that outlaws the capture or killing of any shark species in Bali provincial waters. The creation of a Bali shark sanctuary will be well-received by the international press at a time when Bali is increasingly criticized for its environmental problems, and will prevent even further criticism when information on the thresher shark slaughter is exposed internationally. Moreover, such a move would keep Bali in good stead with its competitor destinations for marine tourism, as many of these (including the Maldives, Palau, Micronesia, the Bahamas, and Guam) have recently declared shark sanctuaries to strong international praise. In October 2011 alone, the Marshall Islands created the world's largest shark sanctuary at 1,990,530 km^2. Bali would be well-served to follow suit, noting that a shark sanctuary will not only create a strong positive media impression of the political will to act decisively on serious environmental problems, it will also over time (as shark populations recover) contribute significantly to increasing the value of Bali's marine tourism.

- The last chapter also analyses secondary data on other marine mega fauna around the island (including whales, dolphins, sea turtles and manta rays). Important sites for Bali's marine mega fauna are included in the seven MPA candidate sites identified in June 2010. The 2011 Marine Rapid Assessment does not cover all crucial information for the design of an MPA network, such as mangrove distribution and basic oceanographic information. In depth analyses on social, cultural and economic contexts are also excluded from the report. Nevertheless, despite the absence of all required data, the Precautionary Principle dictates the immediate implementation of conservation management.

- Taking the above recommendations into account, we strongly recommend that the following nine regions within Bali be prioritized for development of MPAs (or improvement of MPA management in the case of those regions that have already gazetted MPAs): Peninsula region (Bukit Uluwatu to Nusa Dua), Nusa Penida, Padang Bai-Candidasa, Tulamben-Amed, Buleleng Timur (Tejakula), Buleleng Tengah (Lovina), Buleleng Barat (Pemuteran), Bali Barat National Park (including Menjangan and Secret Bay), and Perancak. Depending on local conditions (oceanographic, political, and cultural), it may be appropriate in each of these regions to consider a single larger MPA, or a series of smaller MPAs; either way, it is important that all nine of these regions be prioritized within the Bali MPA Network. We also note that this recommendation should not preclude the gazettement of additional MPAs in areas we have not recommended as priority areas; new information (including data on factors the MRAP did not consider such as mangrove or seagrass distribution, etc) may strongly recommend this, or local communities may simply show strong motivation to implement an MPA.

- It is imperative that the Bali government and all stakeholders recognize that effective management of the MPA network will require serious enforcement efforts and will be a relatively expensive undertaking that will need significant governmental funding to succeed. The government should strongly consider working with the marine tourism sector to develop MPA user fee systems (such as those already working effectively in MPAs like Bunaken and Raja Ampat) that could contribute significantly to the costs of enforcement and MPA management. The government should also consider allocating a percentage of tax revenues from both the tourism and fisheries sectors towards management of the MPA network.

- Bali's coastal zone faces a serious problem of marine litter (especially plastic trash) and pollution from high concentrations of nutrients and sewage entering the ocean through streams and rivers, and seepage in areas of large coastal tourism infrastructure development. The Governor's goal to eliminate the use of fertilizers and pesticides in Bali's agriculture by 2014 is highly commendable and will certainly have a positive effect on this problem. However, much more needs to be done in this regard, including a strong public education campaign (backed up with enforcement and fines) to stop the widespread practice of littering and especially waste-dumping in waterways (all of which eventually lead to the sea). Efforts to seriously reduce the amount of plastic packaging from retail outlets (such as a ban on plastic bags) should also be strongly considered.

1. INTRODUCTION

Bali's coasts and sea support an array of very productive ecosystems that provide goods and services to its communities. The dominant tourism industry also contributes to the economic growth and prosperity of these communities. However, this industry has simultaneously created conflicts of interest between stakeholders and resource users. Uncontrolled coastal development has also triggered environmental degradation in Bali.

Anticipating the various impacts of Bali's development, the provincial government has worked hard to create long-term management strategies, for example, through the production of a spatial plan (Local Regulation No. 16/2009). An important component of this initiative is to design and implement Marine Protected Areas (MPAs) and a network of MPAs in Bali, focusing on sustainable and compatible economic activities (e.g., marine tourism, mariculture and sustainable artisanal fisheries).

The Bali provincial government conducted a stakeholder workshop in June 2010 to kick-start the establishment of MPAs and a network of MPAs on the island. The workshop was organised by the Bali Marine and Fisheries Agency, the Nature Conservation Agency, the University of Warmadewa, the University of Udayana, the United States Agency for International Development (USAID), Conservation International (CI) Indonesia, and the Bali Sea Partnership. It was attended by approximately 70 participants from the provincial and regency level government agencies, universities, NGOs, private sectors, and community groups who work on Bali's coastal issues.

The workshop resulted in several important recommendations, for example, the identification of 25 priority sites to consider in the development of Bali MPAs and an island-wide MPA network. The suggested sites included those with existing management regimes (e.g., the Bali Barat National Park and Nusa Penida MPAs) and other locations without any legal management authorities. Later, seven out of the 25 priority sites were selected as MPA candidate sites. However, the 25 priority sites remain important in the future management of marine ecosystems in Bali.

A comprehensive study on the status of marine resources in Bali was considered important in the development of MPAs around the island. The Marine Rapid Assessment Program (MRAP) is a method to rapidly examine existing marine resources. This MRAP report thus contains basic data and information on the current status of marine resources in Bali to support management and conservation efforts.

1.1 Goals

The goals of Bali Marine Rapid Assessment Program (MRAP) are:

1. To assess the current state of Bali's marine biodiversity (including coral reefs status, hard coral resilience levels, and the status of reef fish, echinoderms and crustaceans)
2. To collect detailed spatial data on biological features to consider in the development of MPAs, MPA zoning and an MPA network; with particular regard to the identification of high value conservation areas for endemic and rare coral cover, reef fish, fish spawning and aggregation sites, coral communities exposed to anthropogenic activities, and other biological features
3. To record and integrate the socio-cultural and economic values of Bali's communities into future MPA management

1.2 Method

Methods used in the 2011 Bali MRAP are methods that have been developed by Conservation International across more than 20 years. They have been used in over 23 countries in the Indian, Pacific and Atlantic Oceans. The methods for coral reef and reef fish are as follows:

A. Reef fish

The reef fish study was led by reef fish expert Dr. Gerald Allen using underwater visual census. The survey was conducted for 60–100 minutes per site. Every observed fish species was recorded using pencil on waterproof paper attached to a clipboard. Scientists conducted the first surveys at a 30–50m water depth before proceeding to shallow water. Most of the time, the surveys were conducted at a 5–12m depth. The number and species of reef fish are usually more abundant at this depth. The team also recorded the substrate type (e.g., rocky, flat reef, drop off, cave, rubble or sand).

B. Hard coral (species diversity and reef status)

The hard coral survey was led by Dr. Lyndon DeVantier, a coral expert of more than 20 years. Surveys were conducted at several dive sites representative of a range of habitat types and environmental conditions (e.g., exposure, slope and depth).

Shallow and deep coral communities were surveyed at the same time at all dive sites. Different depths were surveyed at deeper slopes (usually deeper than 10m), shallow slopes, reef tips and reef flats (usually shallower than 10m).

During the 13 days of the MRAP survey, the team covered 33 sites in an anti-clockwise fashion from the Kutuh Temple in southern Bali to Klatakan in western Bali (Figure 1.2, Table 1.1). The sites were selected based on the recommendations of the June 2010 workshop. Thus, the survey was designed to allow the team to cover as many of the 25 potential sites as possible. Combining the results of the 2011 mainland Bali MRAP (33 sites) with those of the 2008 Nusa Penida MRAP (19 sites), the data in this report cover 52 sites in total: representative information to describe the status of coastal ecosystems in Bali. An exception is made for the western coast of Bali (southeast from Klatakan to Uluwatu); the survey of the western coast was made impossible due to strong currents and waves. Thus, additional efforts need to be made to ensure that this part of the island is surveyed in the future.

Table 1.1. The list of survey sites for the Bali MRAP 29 April to 11 May 2011. The fish survey was not conducted in sites 6, 8 and 28. Site 26 was only surveyed for the reef fish component.

Site No.	Survey date	Location name	Coordinates
1	29 April 11	Terora, Sanur (Grand Mirage)	08° 46.228' S, 115° 13.805' E
2	29 April 11	Glady Willis, Nusa Dua (Grand Mirage)	08° 41.057' S, 115° 16.095' E
3	29 April 11	Sanur Channel	08° 42.625' S, 115° 16.282' E
4	30 April 11	Kutuh Temple, Bukit	08° 50.617' S, 115° 12.336' E
5	30 April 11	Nusa Dua	08° 48.025' S, 115° 14.356' E
6	30 April 11	Melia Bali, Nusa Dua	08° 47.608' S, 115° 14.192' E
7	1 May 11	Batu Tiga-Barat (Gili Mimpang)	08° 31.527' S, 115° 34.519' E
8	1 May 11	Batu Tiga-Timur	08° 31.633' S, 115° 34.585' E
9	1 May 11	Tanjung Jepun (Padang Bai)	08° 31.138' S, 115° 30.619' E
10	2 May 11	Gili Tepekong (Candidasa)	08° 31.885' S, 115° 35.167' E
11	2 May 11	Gili Biaha/Tanjung Pasir Putih	08° 30.270' S, 115° 36.771' E
12	3 May 11	Seraya	08° 26.010' S, 115° 41.274' E
13	3 May 11	Gili Selang-Utara	08° 23.841' S, 115° 42.647' E
14	3 May 11	Gili Selang-Selatan	08° 24.079' S, 115° 42.679' E
15	4 May 11	Bunutan, Amed	08° 20.731' S, 115° 40.826' E
16	4 May 11	Jemeluk, Amed	08° 20.221' S, 115° 39.617' E
17	4 May 11	Kepah, Amed	08° 20.024' S, 115° 39.244' E
18	5 May 11	Batu Klebit, Tulamben	08° 16.696' S, 115° 35.826' E
19	5 May 11	Tukad Abu, Tulamben	08° 17.603' S, 115° 36.599' E
20	6 May 11	Alamanda, Buleleng	08° 08.969' S, 115° 24.733' E
21	6 May 11	Penuktukan, Buleleng	08° 08.270' S, 115° 23.622' E
22	7 May 11	Puri Jati, Lovina	08° 11.032' S, 114° 54.869' E
23	7 May 11	Kalang Anyar, Lovina	08° 11.344' S, 114° 53.841' E
24	8 May 11	Taka Pemuteran	08° 07.775' S, 114° 40.007' E
25	8 May 11	Sumber Kima	08° 06.711' S, 114° 36.451' E
26	9 May 11	Anchor Wreck, Menjangan	08° 05.467' S, 114° 30.131' E
27	9 May 11	Coral Garden, Menjangan	08° 05.485' S, 114° 30.486' E
28	9 May 11	Post 2, Menjangan	08° 05.813' S, 114° 31.608' E
29	10 May 11	Secret Bay, Gilimanuk	08° 10.862' S, 114° 26.544' E
30	10 May 11	Secret Bay Reef -utara, Gilimanuk	08° 09.771' S, 114° 27.116' E
31	11 May 11	Klatakan- Pearl Farm 1	08° 13.911' S, 114° 27.249' E
32	11 May 11	Klatakan-Pearl Farm 2	08° 14.000' S, 114° 27.463' E
33	8 May 11	Pura Pulaki	08° 08.719' S, 114° 40.756' E

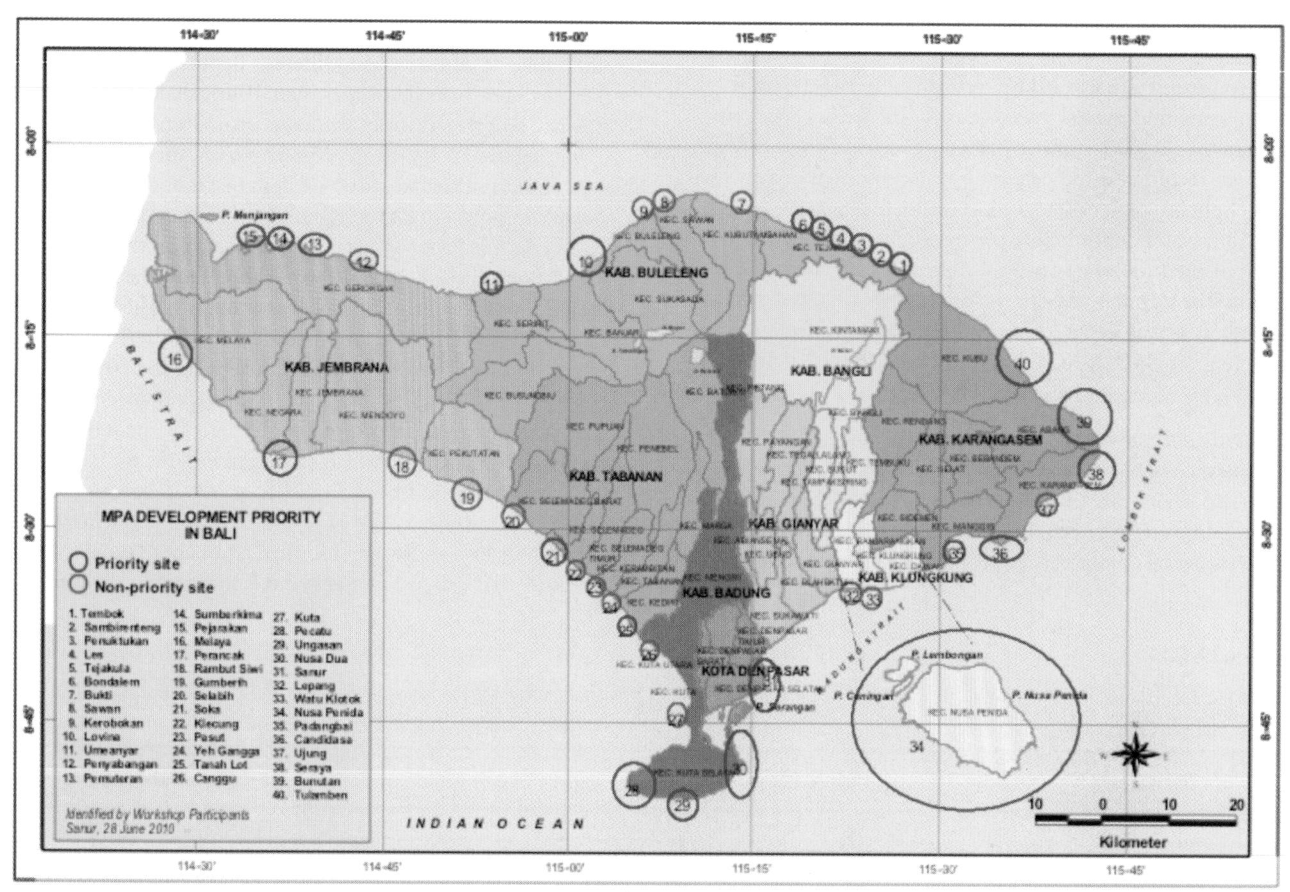

Figure 1.1. Priority sites to be developed as MPAs in Bali (the result of a stakeholder workshop in June 2010)

Figure 1.2. The locations of MRAP in mainland Bali (2011) and Nusa Penida (2008)

2.1 NUSA DUA

Nusa Dua is a 350 ha elite tourism site located at the southernmost corner of Bali. The arid and nonproductive land of the area was acquitted by the government in the 1970s for tourism development. As a result, Nusa Dua has been designed as a comprehensive tourism resort. Built away from the residential area of Bualu village, it offers several places of interest, for example, a water blow and the beaches of Mengiat and Sawangan. Nearby underwater scenery also attracts divers. Nusa Dua is managed by the state company Bali Tourism Development Corporation (BTDC). Several mega hotels, for example, Nikko, Grand Hyatt, Ayodya Resort, Club Med and Nusa Dua Beach are found here.

2.2 SANUR

Sanur stretches from the Padang Galak beach in the north to Merta Sari in the south. A tourist site located near the heart of Denpasar, it also serves as a thoroughfare for traffic heading to Nusa Penida, the Benoa Harbour and Tanjung Benoa. By the 1980s, most Sanur residents were fishers who fished the waters off Sanur and as far as Nusa Dua and Uluwatu. Some of them also traveled to Nusa Penida and Lombok with simple outboard engines and trolling lines. Nowadays, only a small fraction of the villagers fish for daily needs. Most Sanurians are now working in the tourism sector, using their traditional boats (*jukungs*) to take tourists sailing and fishing.

2.3 PADANGBAI

Located at Amuk/Padang Bay, Padangbai is a port that serves as Bali's eastern entrance gate. It harbours ferry boats that connect Bali, Lombok and Nusa Penida, in addition to tour boats that roam the local waters off Padangbai (Goat Island and Blue Lagoon) and Nusa Penida.

The economy of Padangbai depends on tourism and ferry-crossings. It has several tourism sites, for example, the Blue Lagoon and Bias Tugel beaches. Both of these white sand beaches are frequented by foreign tourists, and are popular for being rather isolated and providing degree of privacy. Most tourists came from Europe (Germany) or Asia. Peak season is from July to August and in December.

Although most of the boats in Padangbai are now used for tourism, some of them still used for fishing. These fishing boats are relatively smaller, and some are made of fiberglass. Padangbai fishers still use local waters to catch fish, although they also travel to Nusa Penida and Lombok for this purpose. Sharks are occasionally caught in the triangle section between Padangbai, Gili Tepekong and Nusa Penida.

2.4 CANDIDASA

Candidasa is another tourism site located in Amuk Bay, Karangasem, first established for this purpose around the 1980s. Candidasa is named after a local temple, however due to its close proximity to a large pond, the Candidasa Temple is also often referred to as the Telaga Kauh (the West Lake) Temple.

Despite not being as popular as Sanur and Kuta, Candidasa is a choice for tourists who want to enjoy east Bali's marine and terrestrial tourist attractions. Visitors are usually Europeans (mostly Germans and Dutch). Many Asian tourists also visit the site.

Candidasa offers similar attractions to Padangbai; tourists may enjoy the sea by sailing, fishing, snorkeling or diving. The closeness between two sites means that Candidasa dive operators use the same diving sites as Padangbai's, i.e., Tanjung Jepun, Gili Mimpang, Gili Biaha, Gili Tepekong, as well as Blue Lagoon.

Tourism provides an alternative income for Candidasa fishers. As they finish their daily fishing activities, they will often offer to take tourists fishing, snorkeling or diving, in small boats accommodating up to three customers per trip. An association regulates the tour guide roster.

2.5 SERAYA

The rocky shores of Gili Selang in Seraya are frequently visited by experienced divers due to its amazing underwater scenery and strong, unpredictable currents. No homestays or restaurants are found around Gili Selang despite its popularity as a dive site. Most divers come from Amed, including Bunutan and Jemeluk.

Some residents along the coast of Gili Selang are fishers who operate between 4am and 10am with outboard engines of 8–15 HP. Some also cultivate cassava or raise cows, pigs and goats for extra income. The Gili Selang fishers do not catch sharks, which are rare in the area. Aquarium fishers from Tembok often roam the waters off Gili Selang; travelling by land and camping on the beach. Fishers from other regions also sometimes come by boat and use compressors to catch aquarium fish.

Plate 2.1. Marine tourism provides income for fishers in Candidasa

2.6 AMED

Located in east Bali, Amed is renowned not only for its beautiful underwater scenery but also the Japanese ship that sunk here during World War II. It has several dive sites: Bunutan, Jemeluk and Kepah. Amed's high season is around June to August, with most tourists coming from Europe, although Asians (particularly domestic and Japanese) also frequent this isolated place. Tourists usually come just for diving, thus the general trend is not to stay overnight in Amed.

As with many coastal villagers, the people of Amed are mostly fishers. Some of them only do this on a part-time basis however, occassionally also transporting guests to adjacent tourism sites, or working in restaurants. Tour guide associations regulate the fishers' tour schedule with a roster system. Tourists usually dive, snorkel, fish or sail along the coasts of Amed. Boats from other places also take divers to Jemeluk; these boats have to pay a local village parking fee.

2.7 TULAMBEN

Tulamben is also located in Karangasem, a short distance away from Amed. Tukad Abu Beach is one of the most popular beaches in Tulamben with several villas and restaurants on its shore. Tukad Abu also has several diving sites, for example, Batu Klebit and Batu Belah which offer unique underwater features for photography.

Coastal residents of Tulamben are primarily fishers or tourism workers. The fishers often catch frigate mackerels for sale at the local market (Timbrah) or to villa/restaurant managers. Lately, many of them have been worried by declining and unpredictable fish harvests. Accordingly, some Fish Aggregation Devices (FADs) have been installed to increase harvest rates.

Plate 2.2. Tourist accommodation along the coast of Bunutan, Amed

The fishers of Tukad Abu used to catch sharks in the past by applying long lines approximately 300–500m from the shore. The shark harvest season would take place from the fourth to the fifth month of the Balinese calendar, or the equivalent of August to October. However, shark fishing is no longer conducted due to the various mooring buoys and boats now occupying the "traditional" harvesting area. Tourists who complained about the practice also helped to eliminate this activity.

2.8 TEJAKULA

Tejakula, an administrative district of Buleleng Regency in the north, is a booming marine tourism location. Two famous diving sites here are Alamanda (Gretek Beach) and Penuktukan: two black-sand beaches in Sambirenteng Village. The name "Alamanda" was taken from a local resort and dive operator of the same name. Penuktukan is located close to Alamanda, also in the District of Tejakula.

Not many fishers live around Alamanda. In the 1970s, many villagers became citrus farmers or harvested corals for prestige. When the coral harvesting was prohibited and the citrus crops were attacked by virus in the 1980s, many villagers left their old professions to work instead in construction in Singaraja or Denpasar. In Penuktukan, some villagers are still active fishers to date. They are banded together in a fisher association through which they kite fish frigate mackerel found around Fish Aggregation Devices or catch flying fish with nets.

2.9 SERIRIT

Puri Jati and Kalanganyar are renowned dive sites around Seririt. Most divers are from Asia, particularly Japan, however, Puri Jati is also popular among European and even domestic divers. High season is from June to August. Divers usually visit from, and stay overnight in, Pemuteran or Lovina; despite its fame as an underwater photography spot, not many tourism facilities are available in Puri Jati and Kalanganyar. However, the beach is accessible by motorized vehicles.

Most Puri Jati villagers work as farmers, labourers or private employees. Some also work in the *subak*-style irrigated paddy fields found in the vicinity of Puri Jati. The majority of coastal villagers in the area are not active fishers, and many that do own boats are no longer active. In Kalanganyar, the number of fishers has also been declining; in the 1980s there were hundreds of them, nowadays they number only in the tens. Most Kalanganyar fishers are now farmers and labourers instead.

2.10 PEMUTERAN

Located not far from Pulaki Temple in Gerokgak District, Singaraja, Pemuteran is a booming site equipped with hotels/villas, restaurants and other service providers for visiting tourists. Tourism activities most favoured in Pemuteran are snorkeling and diving. Dive operators offer packages to Pemuteran and Menjangan Island. In addition to beautiful corals and diverse fish, Pemuteran is also famous for its Biorock Technology that uses low electrical current to build artificial reefs to attract fish and divers. Sea turtles are another interesting attraction for Pemuteran. Tourists can view sea turtles directly at a local resort which is also involved in recruiting the local community to participate in the "Sea Turtle Project": where nests found along the coast are relocated to artificial nesting sites. Post-hatching, sea turtle hatchlings are then released back to the sea. The local community, hoteliers and restaurateurs in Pemuteran all participate in firmly suggesting to tourists that they avoid disturbing coral reefs and other marine biota; a strong commitment applauded by many.

Plate 2.3. Local guidance for tourists not to disturb coral reefs in Pemuteran

2.11 MENJANGAN ISLAND

Menjangan Island is located in the Bali Barat National Park. Administratively, it is situated in the Gerokgak District of Buleleng Regency. The underwater beauty off the coast of Menjangan is a magnet for divers. Both divers and tourists reach the island from the ports of Labuan Lalang or Banyu Wedang in the morning and return in the afternoon. Most tourists are Europeans from the Netherlands and France, although Asian tourists (e.g., Japanese and Koreans) also frequent the island.

Traditional fishing still occurs in Menjangan. Fishers come from Java or Bali, operating in the afternoon after the tourists have left the island. When the next morning comes, bringing the tourists back, the fishers disappear one by one.

2.12 GILIMANUK BAY

Gilimanuk Bay is also located in the Bali Barat National Park, although administratively it belongs to the Jembrana Regency. The Bay is shallow (approximately 10 meters), and has two small islands. It is often frequented by international and local tourists, although the latter only come to fish or enjoy the scenery. Muck diving and underwater photography are also often conducted here. The local community offers restaurants, and services and facilities related to diving, for tourists.

2.13 MELAYA

The waters off Melaya are often frequented by sardine fishers. Despite its beauty, Melaya's underwater attractions are still relatively unknown. Visitors are mostly locals who mainly come during holidays to enjoy the coastal scenery.

Pearl farming is an economic activity for Melaya villagers. Many work for a foreign company called "Ocean Blue Pearl Farm" which employs more than 60 locals in its activities, from seeding to harvesting.

Chapter 3

Reef Fishes of Bali, Indonesia

Gerald R. Allen and Mark V. Erdmann

SUMMARY

- A list of fishes was compiled for 29 survey sites. The survey involved approximately 80 hours of scuba diving by G. Allen and M. Erdmann to depths of 70 m.

- A total of 805 species was recorded for the survey.

- Combined with previous survey efforts by the authors, primarily at nearby Nusa Penida in 2008, the current total for the Bali region is 977 species in 320 genera and 88 families.

- A formula for predicting the total reef fish fauna based on the number of species in six key families (Chaetodontidae, Pomacanthidae, Pomacentridae, Labridae, Scaridae, and Acanthuridae) indicates that as many as 1,312 species can be expected to occur in the Bali region.

- Wrasses (Labridae), damselfishes (Pomacentridae), gobies (Gobiidae), cardinalfishes (Apogonidae), groupers (Serranidae), butterflyfishes (Chaetodontidae), and surgeonfishes (Acanthuridae) are the most speciose families on Bali reefs with 114, 96, 84, 59, 54, and 39 species respectively.

- Species numbers at visually sampled sites during the survey ranged from 42 to 248 with an average of 153.

- Sites with the most fish diversity included Anchor Wreck, Menjangan (site 26–248 species), Batu Kelibit, Tulamben (site 18–246 species), Kepa, Amed (site 17 - 230 species), Jemeluk, Amed (site 16–220 species), and Bunutan, Amed (site 15–217 species).

- The majority of Bali fishes have broad distributions in the Indo-Pacific (56.4 %) or western Pacific (25.3 %). Minority categories of special interest involve species that are mainly distributed in the Indian Ocean (3 %) and Indonesian endemics (3.3 %).

- A total of 16 species recorded from Bali are currently known only from the Nusa Tenggara Islands.

- At least thirteen undescribed reef fish species were recorded and collected during the survey, including two fang blennies (*Meiacanthus*), two jawfish (*Opistognathus*), three dottybacks (*Pseudochromis* and *Manonichthys*), a clingfish (*Lepidichthys*), a grubfish (*Parapercis*), a dartfish (*Ptereleotris*), a cardinalfish (*Siphamia*), and two gobiids (*Grallenia* and *Priolepis*). Though most of these undescribed species have been previously recorded from surrounding regions, five were recorded for the first time during the two MRAP surveys.

- Bali is conveniently divisible into four major zones or areas, based on components of the marine fauna in combination with broad-scale physical oceanographic features: Nusa Penida, east coast or Lombok Strait, north coast, and Secret Bay (Gilimanuk).

- Though Bali hosts an astounding diversity of fishes for its size, we also found strong signs of overfishing at nearly every site, with large reef fishes of commercial value nearly absent. Indeed, in over 350 man-hours of diving, the survey team only recorded a grand total of 3 reef sharks (only at Gili Selang and Menjangan), 3 Napoleon wrasse (*Cheilinus undulatus*; observed only at Gili Selang and Tulamben), and 4 coral trout of the genus *Plectropomus*. Equally concerning, the team only recorded a grand total of 5 marine turtles observed during the survey.

- Potential conservation sites based on remarkable fish diversity and excellent habitat conditions include Batu Tiga, Gili Selang, Taka Pemuteran, Sumber Kima, and Secret Bay (Gilimanuk).

3.1 INTRODUCTION

The Indonesia Archipelago is the world's richest region for coral reef fishes (Allen, 2008). Allen and Adrim (2003) provided a comprehensive checklist that included 2057 species. Recent additions (Allen, unpublished data) have since raised the overall total to about 2,250 species. Despite our increasing knowledge of the Indonesian region there is still considerable need for accurate local documentation, particularly for conservation purposes.

Comprehensive documentation of the reef fish fauna based on results of Conservation International's RAP survey during April–May, 2011 are presented in this chapter. The background of this project and detailed descriptions of the survey sites are provided elsewhere in this report. Although this report is focused on the 2011 survey, we have also provided a summary of the combined results of this survey and the 2008 Nusa Penida RAP, with some additional records also included from follow-on dives executed in the months just after the survey.

The principle aim of the fish survey was to provide a comprehensive inventory of the reef fishes of Bali. This segment of the fauna includes fishes living on or near coral reefs to depths of approximately 70 m. Survey activities therefore excluded estuarine species, deepwater fishes and offshore pelagic species such as flyingfishes, tunas, and billfishes.

The results of this survey facilitate a comparison of the reef fish fauna of Bali with other parts of Indonesia as well as other locations in the tropical Indo-Pacific. However, the list of fishes from the survey area is still incomplete, due to the time restriction and the cryptic nature of many small reef species.

3.2 METHODS

The current survey involved a combined total of approximately 80 hours of scuba diving by G. Allen and M. Erdmann to a maximum depth of 70 m. A comprehensive list of fishes was compiled for 29 sites (Appendix 3.1) between 29 April and 11 May 2011. The basic method consisted of underwater observations made during a single dive (rarely two dives) at each site with an average single dive duration of about 80 minutes. The name of each observed species was recorded in pencil on waterproof paper attached to a clipboard. The technique usually involved rapid descent to 30–70 m, then a slow, meandering ascent back to the shallows. The majority of time was spent in the 2–15 m depth zone, which consistently contains the highest number of species. Each dive included all major bottom types and habitat situations in the immediate vicinity.

Fishes were photographed underwater while scuba diving with a Nikon digital SLR camera and 105 mm lens in aluminium housing. Photographs were obtained of approximately 200 species.

Visual surveys were supplemented by collections of mainly cryptic species with the use of clove-oil, rotenone, and spear. Both chemical substances were used in small amounts. Cryptic gobies and other secretive fishes were individually targeted with a clove oil-alcohol mixture by squirting this chemical into caves and crevices. Rotenone was employed primarily in caves or under overhangs, or in some cases along the lower edge of slopes near the coral and sand/rubble interface.

3.3 SURVEY RESULTS

A total of 805 species were collected during the present survey (Lampiran 3.1). Combined with our results of 2008 for Nusa Penida and the first author's previous records of Bali fishes, the reef fish fauna of the Bali region includes 977 species belonging to 320 genera and 88 families. Allen (1997), Kuiter and Tonozuka (2001), and Allen et al (2007) provided illustrations for the majority of species. In addition, detailed coverage of all species is provided by the present authors in their recently-published *Reef Fishes of the East Indies* (Allen and Erdmann, 2012).

3.3.1 ANALYSIS OF SITE DATA

The number of species found at each site is indicated in Table 3.1. The number of species at each site ranged from 42 to 248, with an average of 153 species per site.

Coral and rocky reefs were by far the richest habitat in terms of fish biodiversity. The best sites for fishes (Table 3.2) were invariably locations containing a mixture of substrates including scleractinian corals, soft corals, and rock with algae, seawhips, gorgonians, and sponges. Strong currents were also a contributing factor to species enrichment, particularly the numerous species that feed predominately on current-borne zooplankton. Areas dominated by sand, silt, or rubble substrates were comparatively poor for fishes.

Table 3.1. Number of species observed at each site (note: fishes were not surveyed at sites 6, 8 and 27).

Site	Species	Site	Species	Site	Species
1	96	13	197	23	56
2	162	14	190	24	191
3	157	15	217	25	171
4	91	16	220	26	248
5	131	17	230	28	212
7	187	18	246	29	109
9	115	19	189	30	85
10	183	20	99	31	113
11	143	21	114	32	139
12	117	22	42		

Table 3.2. Richest sites for fishes during 2011 Bali survey.

Site No.	Location	Total fish spp.
26	Anchor Wreck, Menjangan	248
18	Batu Kelit, Tulamben	246
17	Kepa, Amed	230
16	Jemeluk, Amed	220
15	Bunutan, Amed	217
28	Pos 2, Menjangan	212

3.3.2 Coral Fish Diversity Index (CFDI)

In response to the need for a convenient method of assessing and comparing overall coral reef fish diversity between areas in the Indo-Pacific region the first author (see Allen and Werner, 2002) has devised a rating system based on the number of species present belonging to the following six families: Chaetodontidae, Pomacanthidae, Pomacentridae, Labridae, Scaridae, and Acanthuridae. These families are particularly good indicators of overall fish diversity for the following reasons:

- They are taxonomically well documented.
- They are conspicuous diurnal fishes that are relatively easy to identify underwater.
- They include the "core" reef species, which more than any other fishes characterize the fauna of a particular locality. Collectively, they usually comprise more than 50 percent of the observable fishes.
- The families, with the exception of Pomacanthidae, are consistently among the 10 most speciose groups of reef fishes inhabiting a particular locality in the tropical Indo-west Pacific region.
- Labridae and Pomacentridae in particular are very speciose and utilize a wide range of associated habitats in addition to coral-rich areas.

The method of assessment consists simply of counting the total number of species present in each of the six families. It is applicable at several levels:

- single dive sites
- relatively restricted localities (e.g. Bali)
- countries, major island groups, or large regions (e.g. Indonesia)

CFDI values can be used to make a reasonably accurate estimate of the total coral reef fish fauna of a particular locality by means of regression formulas. The latter were obtained after analysis of 35 Indo-Pacific locations for which reliable, comprehensive species lists exist. The data were first divided into two groups: those from relatively restricted localities (reefs and adjacent seas encompassing less than 2,000 km^2) and those from larger areas (reefs and adjacent seas encompassing more than 2,000 km^2). Simple regression analysis revealed a highly significant difference ($P=0.0001$) between these two groups. Therefore, the data were separated and subjected to additional analysis. The Macintosh program Statview was used to perform simple linear regression analyses on each data set in order to determine a predictor formula, using CFDI as the predictor variable (x) for estimating the independent variable (y) or total coral reef fish fauna. The resultant formulae were obtained: 1. total fauna of areas with surrounding seas encompassing more than 2,000 km^2=4.234(CFDI) - 114.446 (d.f=15; $R^2=0.964$; $P=0.0001$); 2. total fauna of areas with surrounding seas encompassing less than 2,000 km^2=3.39 (CFDI) - 20.595 (d.f=18; $R^2=0.96$; $P=0.0001$).

CFDI is useful for short term surveys such as the present one because it is capable of accurately predicting the overall faunal total. The main premise of the CFDI method is that short term surveys of only 15–20 days duration are sufficient to record most members of the six indicator families due to their conspicuous nature. The CFDI for the Bali/Nusa Penida region is 337, composed of the following elements: Chaetodontidae (43), Pomacanthidae (21), Pomacentridae (96), Labridae (114), Scaridae (24), and Acanthuridae (39). The resultant predicted faunal total is 1,312 species. Comparison of this total with the actual number of species (977) currently recorded from the region indicates that at least 335 additional species of shallow reef fishes can be expected. This total includes many species that are not readily recorded by visual methods and small collections. Moray eels (Muraenidae), for example, are notoriously difficult to survey without the use of large quantities of rotenone (a chemical ichthyocide). Only 15 were seen during the surveys reported herein, but on the basis of expected distributions (Allen, unpublished data) at least 35 species should occur in the Bali region. The CFDI method is especially useful when time is limited and there is heavy reliance on visual observations, as was the case for the present survey. The CFDI total indicates that about 75 percent of the fauna was actually recorded during the combined 2008 (Nusa Penida) and 2011 (Bali) surveys.

Table 3.3 presents a comparison of Bali with other Indonesian sites and various Indo-west and central Pacific locations that were surveyed by the author or various colleagues. The Bali/Nusa Penida CFDI value is exceeded only by The Raja Ampat Islands, which is indicative of its impressive reef fish diversity.

3.3.3 Analysis of the Bali reef fish fauna

The most abundant families in terms of number of species are wrasses (Labridae), damselfishes (Pomacentridae), gobies (Gobiidae), cardinalfishes (Apogonidae), groupers (Serranidae), butterflyfishes (Chaetodontidae), surgeonfishes (Acanthuridae), parrotfishes (Scaridae), and snappers (Lutjanidae). These 10 families collectively account for about 59 percent of the total reef fauna (Table 3.4).

The relative abundance of Bali fish families is very similar to that found at other Indo-Pacific locations. Labridae, Pomacentridae, and Gobiidae are typically the most speciose families, although the order of these groups is variable according to location. The gobiidae is frequently the most abundant, which is not surprising given that approximately 600 species inhabit Indo-Pacific coral reefs. This family no doubt contains more species than any other family at Bali as well, but it is difficult to comprehensively survey the group due to the very small size and cryptic nature of many species. Also their proclivity for open sand and rubble habitats runs counter to the RAP survey method, which focuses mainly on reef substrate.

3.3.4 Zoogeographic affinities

Bali belongs to the Western Pacific faunal community, which forms an integral part of the greater Indo-West and central Pacific biotic province. Its reef fishes are very similar to those

Table 3.3. Coral fish diversity index (CFDI) values for restricted localities, number of coral reef fish species as determined by surveys to date, and estimated numbers using the CFDI regression formula (refer to text for details).

Locality	CFDI	No. reef fishes	Estim. reef fishes
Raja Ampat Islands, West Papua, Indonesia	373	1437	1465
Bali and Nusa Penida	**337**	**977**	**1,312**
Maumere Bay, Flores, Indonesia	333	1111	1108
Milne Bay Province, Papua New Guinea	333	1109	1295
Halmahera, Indonesia	327	974	1271
Berau, East Kalimantan, Indonesia	316	875	1050
Togean and Banggai Islands, Indonesia	308	819	1190
Cendrawasih Bay, West Papua, Indonesia	302	965	1165
Solomon Islands	301	1019	1160
Northern Tip of Palawan, Philippines	292	1003	1122
Komodo Islands, Indonesia	280	750	928
Yap State, Micronesia	280	787	928
Verde Passage, Philippines	278	808	921
Madang, Papua New Guinea	257	787	850
Kimbe Bay, Papua New Guinea	254	687	840
Manado, Sulawesi, Indonesia	249	624	823
Capricorn Group, Great Barrier Reef	232	803	765
Chuuk State, Micronesia	230	615	759
Brunei, Darussalam	230	673	759
Ashmore/Cartier Reefs, Timor Sea	225	669	742

Locality	CFDI	No. reef fishes	Estim. reef fishes
Kashiwa-Jima Island, Japan	224	768	738
Samoa Islands	211	852	694
Chesterfield Islands, Coral Sea	210	699	691
Pohnpei and nearby atolls, Micronesia	202	470	664
Layang Layang Atoll, Malaysia	202	458	664
Bodgaya Islands, Sabah, Malaysia	197	516	647
Pulau Weh, Sumatra, Indonesia	196	533	644
Izu Islands, Japan	190	464	623
Christmas Island, Indian Ocean	185	560	606
Sipadan Island, Sabah, Malaysia	184	492	603
Rowley Shoals, Western Australia	176	505	576
Cocos-Keeling Atoll, Indian Ocean	167	528	545
North-West Cape, Western Australia	164	527	535
Tunku Abdul Rahman Is., Sabah	139	357	450
Lord Howe Island, Australia	139	395	450
Monte Bello Islands, W. Australia	119	447	382
Bintan Island, Indonesia	97	304	308
Kimberley Coast, Western Australia	89	367	281
Johnston Island, Central Pacific	78	227	243
Midway Atoll	77	250	240
Norfolk Island	72	220	223

Table 3.4. Largest fish families at Bali.

Rank	Family	Species	% of total species
1	Labridae	114	11.7
2	Pomacentridae	96	9.8
3	Gobiidae	84	8.6
4	Apogonidae	59	6.0
5	Serranidae	54	5.5
6	Chaetodontidae	43	4.4
7	Acanthuridae	39	4.0
8	Blenniidae	27	2.8
9	Scaridae	24	2.5
10	Lutjanidae	22	2.3

Table 3.5. Zoogeographic analysis of Bali reef fishes. Each category is mutually exclusive.

Distribution category	No. Spp.	% of fauna
Indo-West Pacific	551	56.39
Western Pacific	247	25.28
Indo-Australian Archipelago	87	8.90
Indonesian endemic	32	3.27
Indian Ocean	29	2.97
Undetermined	19	1.94
Circumtropical	7	0.07
Japan and Nusa Penida	5	0.05

Table 3.6. Indian Ocean species occurring at Bali.

Family Caesionidae	Labridae
Caesio xanthonota	*Bodianus diana*
Family Mullidae	*Gomphosus caeruleus*
Parupeneus macronemus	*Halichoeres chrysotaenia*
Parupenus trifasciatus	*Leptojulis cyanotaenia*
Family Chaetodontidae	**Family Scaridae**
Chaetodon collare	*Chlorurus capistratoides*
Chaetodon decussates	**Family Blenniidae**
Chaetodon guttatissimus	*Entomacrodus vermiculatus*
Chaetodon trifasciatus	**Family Gobiidae**
Family Pomacanthidae	*Trimma fucatum*
Centropyge eibli	**Family Acanthuridae**
Genicanthus caudivittatus	*Acanthurus leucosternon*
Family Pomacentridae	*Acanthurus tennentii*
Amphiprion akallopisos	*Acanthurus tristis*
Amphiprion sebae	*Ctenochaetus truncatus*
Chromis dimidiate	*Naso elegans*
Chromis opercularis	**Family Balistidae**
Pomacentrus alleni	*Melichthys indicus*

inhabiting other areas within this vast region that stretches from East Africa and the Red Sea to the islands of Micronesia and Polynesia. Although most families and many genera and species are consistently present across the region, the species composition varies greatly according to locality.

Dispersal capabilities and the larval lifespan of a given species are usually reflected in its geographic distribution. Most reef fishes have a relatively long pelagic stage, hence a disproportionate number of wide-ranging species inhabit tropical seas. This is clearly demonstrated in the Bali community with approximately 56% of the species exhibiting distribution patterns that encompass much of the Indo-west and central Pacific region. Many species range from East Africa to either the western edge of the Pacific or well eastward to Micronesia and Polynesia.

Table 3.5 presents the major zoogeographic categories of Bali reef fishes. In addition to wide ranging Indo-Pacific species other major categories include species (about 25%) that are widely distributed in the western Pacific and species (about 9%) that are largely restricted to the Indo-Australian Archipelago (Andaman Sea eastward to the Melanesian Archipelago and Australia northward to the Philippines).

A total of 29 species have distributions that are mainly confined to the Indian Ocean (Table 3.6). Many of these range widely in the Indian Ocean, occurring as far westward as the East African coast and Red Sea. However, a few such as *Centropyge eibli* and *Pomacentrus alleni* are confined to the eastern portion of the Indian Ocean. In most cases the Bali region represents the eastern limit of distribution. Several examples of the Indian Ocean species occurring at Bali are illustrated in Plate 3.1.

Seven species (*Rhincodon typus, Manta birostris, Echeneis naucrates, Thunnus albacares, Melichthys niger, Diodon hystrix,* and *Mola mola*) exhibit circumtropical distributions. These are primarily species that have lengthy pelagic larval stages and settle on reefs at a relatively large size (e.g. *Melichthys*) or are adapted for life in the pelagic realm, far from shore (e.g. *Rhincodon, Manta, Thunnus,* and *Mola*). The sharksucker, *Echeneis*, is readily dispersed throughout the world's tropical seas by a variety of host organisms including large pelagic fishes, marine mammals, and turtles.

Five species of Bali/Nusa Penida fishes, including the wobbegong shark *Orectolobus japonicus*, apogonid *Apogon schlegeli*, scorpaenid *Scorpaenodes evides*, pomacentrid *Chromis albicauda,* and gobiid *Trimma imaii* display an unusual disjunct distribution that involves Japan and the Bali region. Most likely these species were once widely distributed in cooler waters of the west Pacific north of Indonesia, but warming of surface waters have caused widespread extinction. These species apparently now persist as "relict" populations in subtropical waters of Japan and Bali, where cool upwelling results in lowered sea temperatures. Although these species are currently known in Indonesia only from Bali and Nusa Penida, they can be expected in other parts of the Lesser Sunda Islands that are exposed to cool upwelling.

Zoogeographically, the most interesting groups of Bali fishes are those that exhibit highly restricted distribution patterns including the 32 Indonesian endemics, but especially the 16 species currently known only from the Lesser Sunda Islands (Table 3.7 and Plate 3.2). Allen & Adrim (2003) and Allen and Erdmann (2012) documented the highly endemic nature of the Indonesian reef fish fauna, indicating that the Nusa Tenggara Islands (i.e. Lesser Sunda Islands) is the richest area for regional endemism in Indonesia and the East Indian Region in general. Intensive work in the Bird's Head Peninsula in West Papua reveals that area as the second richest location for endemic species in Indonesia. The evolution of Bird's Head endemics has been fuelled by a combination of rich habitat diversity, tectonic activity, and sea level fluctuations, whereas the Lesser Sunda endemics appear to have resulted from the unique habitat conditions along the southern exit of the "Indonesian Through flow", an area of strong currents and cold upwelling.

Table 3.7. Lesser Sunda endemic reef fishes occurring at Bali.

Family	Species	Geographic distribution
Pseudochromidae	*Haliophis aethiopus*	Bali and Nusa Penida
	Pseudochromis aurulentus	Nusa Penida and Komodo
	Pseudochromis oligochrysus	Bali to Alor
	Pseudochromis rutilus	Nusa Penida
	Pseudochromis steenei	Bali to Alor
	Manonichthys sp.	Bali to Komodo
Apogonidae	*Apogon lineomaculus*	Bali to Komodo
	Siphamia sp.	Bali
Pomacentridae	*Chromis pura*	Nusa Penida and Alor
	Chromis sp.	Nusa Penida
Tripterygiidae	*Helcogramma kranos*	Bali to Komodo
	Helcogramma randalli	Bali to Alor
Bleniidae	*Meiacanthus cyanopterus*	Bali to Alor
	Meiacanthus abruptus	Bali to Komodo
Gobiidae	*Grallenia baliensis*	Bali
Acanthuridae	*Prionurus chrysurus*	Nusa Penida to Komodo

3.3.5 Geminate species and hybridization

Randall (1998) gave examples of 52 species pairs that involve closely related Indian Ocean and Pacific Ocean species. He suggested that these "geminate" species had evolved as the result of a common event—the former widely distributed Indo-Pacific ancestral species stock having been divided by lowered sea levels in the past that resulted in an East Indian barrier. For example during the Pleistocene, this barrier would have consisted of dry land extending from the northern tip of Sumatra to Timor, with only one small opening between Bali and the Lesser Sunda Islands.

One of the unusual features of the Bali fish fauna is the presence of both Indian Ocean and Pacific Ocean members of several of these twin-species pairs (Table 3.8 and Plate 3.3). In nearly every case the Pacific Ocean member of the pair is the most common in comparison with its relatively rare Indian Ocean counterpart. This phenomenon possibly indicates predominately southward flowing currents.

Hybridization is a relatively rare phenomenon in marine fishes compared to their freshwater counterparts. However, tropical butterflyfishes (Chaetodontidae) and angelfishes (Pomacanthidae) are exceptions with many hybrids having been reported. Pyle and Randall (1994) provided references to 15 butterflyfish hybrids and noted that at least 12 others remain to be documented in the literature. These authors also documented 11 examples of probable hybridization in angelfishes. Moreover, a recent study by Hobbs et al (2008) reported 11 hybrids belonging to six families at Christmas Island, which lies approximately 1000 km southwest of Bali or 350 km directly south of Genteng Point, west Java.

No hybrids were observed during the current Bali RAP, but were seen on several occasions during the 2008 survey

Table 3.8. Examples of geminate species pairs recorded at Bali.

Family	Pacific Ocean Species	Indian Ocean Species
Caesionidae	*Caesio teres*	*Caesio xanthonota*
Chaetodontidae	*Chaetodon vagabundus*	*Chaetodon decussatus*
	Chaetodon punctatofasciatus	*Chaetodon guttatissimus*
	Chaetodon lunulatus	*Chaetodon trifasciatus*
Pomacanthidae	*Centropyge vroliki*	*Centropyge eibli*
Pomacentridae		
	Chromis margaritifer	*Chromis dimidiata*
	Chromis xanthurus	*Chromis opercularis*
	Pomacentrus coelestis	*Pomacentrus alleni*
Scaridae	*Chlorurus bleekeri*	*Chlorurus capistratoides*
Acanthuridae	*Acanthurus pyroferus*	*Acanthurus tristis*
	Ctenochaetus cyanocheilus	*Ctenochaetus truncatus*
	Naso lituratus	*Naso elegans*

Plate 3.1. Example of Indian Ocean species at Bali (from left to right starting on top): *Acanthurus tristis, Amphiprion sebae, Chaetodon trifasciatus, Chromis opercularis, Leptojulis chrysotaenia,* and *Melichthys indicus.*

Plate 3.2. *Apogon lineomaculus,* 6 cm total length. Known only from Bali and Komodo.

Plate 3.3. Examples of geminate species pairs (Indian Ocean species on left and Pacific species on right): upper–*Chaetodon decussatus* and *C. vagabundus*; middle–*Chromis dimidiata* and *C. margaritifer*; lower–*Ctenochaetus cyanocheilus* and *C. truncatus.*

Plate 3.4. An example of hybridization (center) at Nusa Penida between *Centropyge eibli* (left) and *C. vroliki* (right).

Plate 3.5. Examples of Bali species associated with areas of cool upwelling: from left to right – *Prionurus chrysurus, Springeratus xanthosoma,* and *Mola mola.*

Plate 3.6. *Parapercis bimacula*, 11 cm total length.

Plate 3.7. *Manonichthys* sp., 3.5 cm total length.

Plate 3.8. Two new *Pseudochromis* from Bali and Nusa Penida, 7 cm total length.

Plate 3.9. *Siphamia* sp., 3.5 cm total length.

Plate 3.10. Two new jawfishes (Opistognathidae) from Bali (left to right): *Opistognathus* species 1, 4 cm total length, *Opistognathus* species 2, 3.5 cm total length.

Plate 3.11. *Meiacanthus abruptus*, 7 cm total length.

Plate 3.12. *Meiacanthus cyanopterus*, 6 cm total length.

Plate 3.13. *Priolepis* sp., 2.5 cm total length.

Plate 3.14. *Grallenia baliensis*, 2.5 cm total length.

Plate 3.15. *Lepadichthys* sp., 3 cm total length.

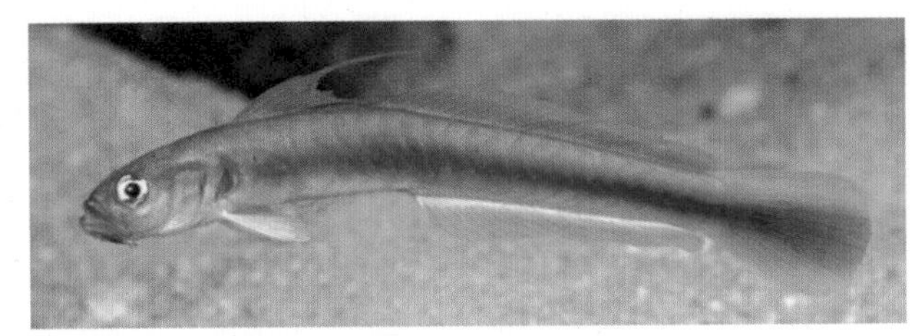

Plate 3.16. *Ptereleotris rubristigma*, 10 cm total length.

Plate 3.17. New distributional records included (from left to right): *Chaetodon reticulatus, Abudefduf lorentzi,* and *Cirrhilabrus pylei.*

Plate 3.18. The introduced Banggai Cardinalfish (*Pterapogon kauderni*), 8 cm TL, Secret Bay, Bali.

Figure 3.1. Satellite image of Secret Bay at Gilimanuk.

of Nusa Penida. These cases involved crosses between the butterflyfishes *Chaetodon guttatissimus* and *C. punctofasciatus* and the angelfishes *Centropyge eibli* and *C. vroliki* (Plate 3.4). Although no hybrids were detected between the closely related *Chaetodon lunulatus* and *C. trifasciatus*, several mixed pairs between these species were seen during both the 2008 and 2011 surveys.

3.3.6 Cool upwelling

The eastern coast of Bali, including the Lombok Strait and the island of Nusa Penida, is characterized by swift currents and cool temperatures due to upwelling of deep water. It is not unusual to encounter temperatures in the low 20s or even colder. Table 3.9 presents a list of species that are frequently associated with these areas of upwelling and three examples are also shown in Plate 3.5.

Table 3.9. Cool upwelling associated species occurring at Bali.

Family Chaetodontidae	Family Clinidae
Chaetodon guentheri	*Springeratus xanthosoma*
Heniochus diphreutes	**Family Acanthuridae**
Family Pomacanthidae	*Prionurus chrysurus*
Chaetodontoplus melanosoma	**Molidae**
Family Pomacentridae	*Mola mola*
Chromis albicauda	
Chromis pura	

3.3.7 New species

Several undescribed species were recorded during the current RAP survey. These are discussed in more detail in the following paragraphs, noting that many of these new species have since been described by the authors and specialist colleagues (e.g., Allen and Erdmann, 2012; Smith-Vaniz and Allen, 2011; Gill, Allen and Erdmann, 2012).

Parapercis bimacula Allen and Erdmann, 2012 (Pinguipedidae; Plate 3.6)—This beautiful grubfish has now been recorded from Bali, Komodo, Pulau Weh (Sumatra) and the Andaman Islands, where it is found typically on sand/rubble bottoms with scattered live coral rock outcrops in 2–8 depth. It was recently described by the authors in their book in reef fishes of the East Indies.

Manonichthys species (Pseudochromidae; Plate 3.7)—This species was observed and photographed in 29–30 m depth at two sites (25 and 28) on the northwestern coast, including Menjangan Island. It is also known from Komodo. It is closely related to *M. alleni* of northern Borneo and is currently being studied by pseudochromid expert Anthony Gill of Sydney University, Australia, who will determine its definitive status.

Pseudochromis oligochrysus Gill, Allen and Erdmann, 2012 (Pseudochromidae; Plate 3.8, left)—This species was not encountered during the present survey, but several specimens were collected during the 2008 RAP at Nusa Penida. It is locally common on moderate slopes at depths between about 25–50 m. The species was recently described in early 2012 by Anthony Gill in conjunction with the authors.

Pseudochromis rutilus Gill, Allen and Erdmann, 2012 (Pseudochromidae; Plate 3.8, right)—This new species (Plate 3.7) was collected at Nusa Penida in 2008 and also during the current survey at Menjangan (site 27) at depths between about 60–70 m. It is generally seen around rocky outcrops, sponges and in crevices on outer reef slopes. It has also been collected in the Alor region of Nusa Tenggara. As with the previous species, this one was described in January 2012 by Anthony Gill and the authors.

Siphamia species (Apogonidae; Plate 3.9)—A single specimen of this unusual cardinalfish was collected at Menjangan Island (site 27) in 70 m depth. Like all members of the genus *Siphamia* it is characterized by a silvery bioluminescent organ along the lower side of the body. It appears to belong to an undescribed species closely related to the poorly-known species *S. argentea*, and is distinguished by its unique colouration in combination with a striated light organ, relatively deep body, and complete lateral line. The species is currently being examined closely by South African apogonid expert Ofer Gon.

Opistognathus species 1 (Opistognathidae; Plate 3.10, left)—Jawfishes have been a particularly rewarding group with many new discoveries by the authors in recent years. This species remains undescribed, but has been previously reported from the Andaman Islands, Kalimantan (Derawan), Philippines (Siquijor Island), and Indonesia (Morotai Island and Cenderawasih Bay, West Papua). It inhabits Sand/rubble bottoms near reefs in 20–70 m. Three specimens were collected during the present survey at site 25 (Sumber Kima). It will soon be described by American opistognathid expert, William Smith-Vaniz.

Opistognathus species 2 (Opistognathidae; Plate 3.10)—This new jawfish has previously been collected at Brunei and the Philippines. It inhabits turbid coastal reefs on sand/rubble bottoms near reefs in areas of periodic strong currents in about 15-70 m depth. A single specimen was collected during the Bali survey at site 25 (Sumber Kima). The species will also be described by William Smith-Vaniz.

Meiacanthus abruptus Smith-Vaniz and Allen, 2011 (Blenniidae; Plate 3.11)—This new species was first collected by G. Allen at Komodo in 1995. About 10 individuals were photographed at Secret Bay, Gilimanuk (site 30) during the current Bali survey. It was found on a small patch reef with nearly 100 percent coral cover in 2–4 m depth. The species is characterized by a yellow head and pair of black stripes on the body. The species has now been described as of October 2011 (Smith-Vaniz and Allen, 2011).

Meiacanthus cyanopterus Smith-Vaniz and Allen, 2011 (Blenniidae; Plate 3.12)—A single specimen of this deep-dwelling species was observed in 70 m depth at site 19. It is currently known only from the Nusa Tenggara Islands at Bali and the Alor region. Like the former species, this species has now been described as of October 2011 (Smith-Vaniz and Allen, 2011).

Priolepis species (Gobiidae; Plate 3.13)—This apparently undescribed species superficially resembles *P. pallidicincta* Winterbottom & Burridge, but markedly differs in having transverse rows of cheek papillae. It was collected at two sites (10 and 26) during the present survey at a depth of 70 m.

Grallenia baliensis Allen and Erdmann, 2012 (Gobiidae; Plate 3.14)—This diminutive (maximum size about 2.5 cm TL) goby was previously known on the basis of a few specimens collected in the Tulamben area on sand/pebble bottoms in 5–15 m depth. It was found at Amed (site 17) and Buleleng (site 21) during the current survey. Detailed examination reveals it is a new species distinguished by its unique colour pattern, lack of filamentous dorsal spines in males, relatively short anal and second dorsal fin rays, and a short pectoral fin. The species was recently described by the authors in their book on reef fishes of the East Indies in March 2012.

Lepadichthys species (Gobiesocidae; Plate 3.15)—This apparently undescribed species was previously known only on the basis of underwater photographs from Flores, Indonesia and Manus Island, Papua New Guinea. It is normally dark reddish brown with distinctive white stripes, one midlaterally on each side, and another mid-dorsally from the top of the snout to the upper edge of the caudal fin. The species sometimes seeks shelter among the spines of *Diadema* urchins and generally occurs in 5–15 m depth. A single specimen was collected during the 2011 survey at site 25 (Sumber Kima).

Ptereleotris rubristigma Allen, Erdmann and Cahyani, 2012 (Ptereleotridae; Plate 3.16)—This species was previously misidentified as *P. hanae*, but differs from that species in lacking elongate filaments on the caudal fin, in having a filamentous second dorsal spine in adult males, and a reddish mark (sometimes absent) on the pectoral-fin base. It is widely distributed in Indonesia and surrounding regions. During the current survey it was observed at Seraya (site 12), Amed (site 16), and Taka Pemuteran (site 24). The habitat consists of open sand and rubble bottoms at depths between about 5–50 m. The species was recently described by the authors (and geneticist colleague Dita Cahyani) in their book on reef fishes of the East Indies in March 2012.

3.3.8 Range extensions and noteworthy records

Chaetodon reticulatus Cuvier, 1831 (Chaetodontidae; Plate 3.17, left)—This species is widely distributed in the western Pacific, mainly at islands of Oceania eastward to the Line and Society islands. It has only been reported from Indonesia at Halmahera and off northern Sulawesi and our current Bali record (site) represents a range extension of approximately 1500 km.

Abudefduf lorentzi Hensley & Allen, 1977 (Pomacentride; Plate 3.17, middle)—This is a common shallow water inhabitant of eastern Sulawesi, Halmahera and the Papuan region of Indonesia, as well as Papua New Guinea, Solomon

Islands, and the Philippines. It is generally replaced by the closely related *A. bengalensis* east and south of Sulawesi. Therefore it was unexpected to see a single subadult along the shoreline at site 28 (Menjangan), representing a range extension of approximately 900 km.

Cirrhilabrus pylei. (Labridae; Plate 3.17, right)—Although previously reported from Bali on the basis of underwater photographs we were able to confirm its presence in the region with the collection of specimens at both Nusa Penida in 2008 and site 28 (Menjangan) during the present survey. Most previous records of this spectacular species are from the Melanesian Archipelago, including West Papua, Papua New Guinea, Solomon Islands, and Vanuatu.

3.3.9 Introduced fishes

Although introduced species constitute a relatively minor portion of global fish communities they have the potential to alter the dynamics of local fish populations. The scorpaenid fish *Pterois volitans* is a classic example of this phenomenon. Although it is commonly encountered throughout its western and central Pacific distributional range, it is usually seen in relatively low numbers. For example, it is not unusual to see one per several dives during typical RAP surveys. This species is collected for the aquarium trade and was released in Florida waters approximately 20 years ago. Apparently it is now in "plague proportions" in certain areas of the eastern USA coast and Caribbean Sea and has significantly impacted local fish communities due to its predatory habits, which target a wide range of small fishes and invertebrates. In addition to accidental and deliberate release of aquarium fishes other introductions are the result of intentional stock enhancement of commercially valuable species (e.g. *Lutjanus kasmira* in Hawaii), access to previously separated seas due to canal construction (e.g. Red Sea introductions in the Mediterranean Sea via the Suez Canal), and transport of larvae and small benthic fishes in the ballast tanks of freight ships.

The Banggai Cardinalfish (*Pterapogon kauderni*, Plate 3.18) has a very restricted natural distribution, limited to the Banggai Islands and nearby adjacent areas of central-eastern Sulawesi. This strikingly handsome fish was introduced to the aquarium trade in 1995 and was an instant sensation, selling for approximately $100 per fish during the initial months of its availability. Huge numbers were shipped overseas, primarily via fish sellers based at Bali and north Sulawesi.

Consequently the fish was intentionally released in the Lembeh Straits region of Sulawesi and at Gilimanuk, Bali, where populations continue to flourish. The Bali population appears to be confined to a very small area near the beach on the south side of the entrance to Secret Bay at Gilimanuk. It is associated with *Diadema* sea urchins, which are common in the shallows and around the sunken wreckage of a small boat. We estimate a current population of approximately 1000 individuals, and compared to casual observations made by us about two years ago, the numbers appear to be increasing. There is no indication that the species has penetrated outside of the bay, and due its peculiar reproductive mode (eggs and small young orally incubated by male) and lack of pelagic dispersal capability, the expansion of its range around Bali will likely be a slow process. The species feeds on planktonic organisms and small benthic invertebrates. Therefore, its impact on the general fish population appears to be minimal, possibly limited to a few other apogonid species that compete for living space among the spines of *Diadema.* One positive aspect of this introduction is that tourist divers are attracted to the site by the rare opportunity to photograph this spectacular species, sparing the expense and logistic difficulties of travelling to the Banggai Islands.

3.4 SITES OF SPECIAL SIGNIFICANCE FOR FISHES WITH POTENTIAL CONSERVATION VALUE

Comparison of major geographic areas of the Bali region

The current survey reveals that Bali has a remarkably diverse reef fish fauna, reflecting a relatively wide range of habitat variability. The CFDI method of predicting an overall faunal total, based on key index families indicates that Bali is one of Indonesia's richest areas for reef fishes and therefore globally important with regards to conservation significance. The fish community is particularly impressive considering that sheltered lagoon conditions are generally lacking. Therefore, the species associated with this habitat are either rare or absent. Bali is conveniently divisible into discrete zones or areas, based on components of the marine fauna in combination with broad-scale physical oceanographic features, particularly temperature and currents, including upwelling. A comparison of fish diversity for the major geographic areas is presented in Table 3.10.

The north coast is the richest area for fish diversity. It contains Bali's best examples of coral development as exemplified by reefs at Amed and Menjangan Island. Within this area there are also interesting silt bottom "muck dive" areas that harbor unusual fishes, not usually seen on typical coral reefs.

Nusa Penida is worthy of a separate zone, due to its relative isolation, full exposure to the Indian Ocean, and general habitat conditions typified by swift currents and cold upwelling.

Table 3.10. Comparison of species totals for major geographic areas of the Bali region.

Geographic area	No. spp.	Spp./site
Northern Bali	622	214*
Nusa Penida	573	161
Eastern Bali	510	147
Gilimanuk	153	97
Grand Total	**964**	

* excludes silt bottom sites (20–23) in Lovina area.

The east coast of Bali, consisting of the Lombok Strait forms a third major zone. Like Nusa Penida it is subjected to periodic strong currents and cool upwelling. Several of the "signature"species that are also typical of Nusa Penida, for example the Yellowtail Surgeonfish (*Prionurus chrysurus*) and Ocean Sunfish (*Mola mola*), are also found here.

"Secret Bay" at Gilimanuk forms a fourth major zone. Although occupying an extremely small area (approximately 5.5 km^2), the bay is highly unique in the context of Bali marine habitats and the fish community it supports. The bay is bounded by mangrove and contains a number of patch reefs with good live coral growth as well as extensive silt-bottom habitat that is home to a wealth of unusual fishes not usually seen in other parts of the island.

The south coast was inadequately surveyed to determine if it is deserving of separate major area status. Only two sites (31–32) were sampled. These preliminary observations indicate there may be justification for including it in the same faunal region as eastern Bali.

3.5 CONSERVATION RECOMMENDATIONS

Though Bali hosts an astounding diversity of fishes for its size, we also found strong signs of overfishing at nearly every site. Large reef fishes of commercial value were nearly absent during this most recent survey. Indeed, in over 350 man-hours of diving, the survey team only recorded a grand total of 3 reef sharks (only at Gili Selang and Menjangan), 3 Napoleon wrasse (*Cheilinus undulatus*; observed only at Gili Selang and Tulamben), and 4 coral trout of the genus *Plectropomus*. Equally concerning, the team only recorded a grand total of 5 marine turtles observed during the survey. These disturbingly low numbers should serve as a sharp warning to the government of Bali; one would normally expect to see these numbers on a single dive on a healthy reef rather than as the cumulative total for 33 survey sites!

In order to counter this strong trend towards overfishing on Bali's reefs, it is highly recommended to establish a network of "no take" marine protected areas (MPAs) around the island that contain representative faunal communities in each of the major areas outlined above. The benefits for establishing effective marine protected areas include the sustainability of high biodiversity and increased economic value due to their attractiveness to divers and snorkellers, and also from the well-documented "spillover effect" that results from the biomass buildup in no-take areas and leads directly to increased catches of food fishes in areas adjacent to these MPAs.

The previous (2008) report recommended several sites that are most worthy of full protection at Nusa Penida including Crystal Bay, Toyapakeh, Batu Abah and Teluk Batu Abah based on their respective fish communities and remarkable reef habitat. Using similar criteria we also recommend the following sites for consideration for designation in no-take zones as a result of the 2011 survey.

Batu Tiga near Candidasa— these rocky islets support relatively rich coral communities and associated fish fauna, although large predatory species such as sharks and groupers were absent. A total of 187 species was recorded at West Batu Tiga (site 7), the third highest number for the eastern coast.

Gili Selang, near northeast corner of Bali (sites 13–14)—an area of good microhabitat diversity with a rich coral reef fish assemblage as well as soft-bottom and surge-zone associated species. The two highest totals for the east coast, 197 and 190 species, were recorded at North and South Selang respectively.

Taka Pemuteran and Sumber Kima reef complexes, northwest Bali (sites 24–25)—Both these areas exhibit good microhabitat diversity and support relatively rich fish communities (191 and 171 species respectively). The site at Taka Pemuteran was particularly rich for live corals and associated reef fishes. Both of these reef complexes have excellent potential for "no-take" zonation, with the intent of enriching fisheries in adjacent areas and providing high quality recreational diving.

Secret Bay, Gilimanuk (sites 29–30)—The nearly enclosed lagoon system at Secret Bay is highly unique and supports a wealth of fishes that are either absent or rare at other parts of the island. There is a need for further survey work to establish a comprehensive list of the bay's fish fauna. The bay provides a good blend of silty open bottom habitat, shoreline fringing reefs, and mid-lagoon patch reefs, as well as a mangrove shore and several mangrove-lined islands. Special conservation protection of this unique area is recommended, including protection of adjacent mangrove habitat.

ACKNOWLEDGEMENTS

We thank the honourable Governor of Bali I Made Mangku Pastika and the Bali provincial Department of Marine Affairs and Fisheries for inviting us to conduct the marine biodiversity survey which led to this discovery, and the United States' Agency for International Development's (USAID) Coral Triangle Support Program for funding the survey. We also acknowledge Conservation International's Indonesia Marine Program for organizing and hosting the survey, in particular our colleagues Ketut Sarjana Putra, Made Jaya Ratha, and Muhammad (Erdi) Lazuardi, and we also thank Putu (Icha) Mustika and Made Jaya Ratha for their hard work preparing this RAP report. We furthermore thank Wolcott Henry and the Clark and Edith Munson Foundation and the Paine Family Trust for their support of the first author's taxonomic work. Finally, we thank Michael Cortenbach of Bali Diving Academy and Adam Malec of Scubadamarine for their excellent diving support for our survey.

REFERENCES

Allen, G.R. 1997. *Marine fishes of south-east Asia*. Western Australian Museum: Perth, 292 pp.

Allen, G.R. 2008. Conservation hotspots of biodiversity and endemism for Indo-Pacific coral reef fishes. *Aquatic Conservation: Marine and Freshwater Ecosystems* 18: 541–556.

Allen, G.R. and Adrim, M. 2003. Coral reef fishes of Indonesia. *Zoological Studies* 42(1): 1–72.

Allen, G.R. and Erdmann, M.V. 2012. *Reef Fishes of the East Indies*. Volumes I-III. Tropical Reef Research: Perth, Australia, 1292 pp.

Allen, G., Steene, R., Humann, P., and Deloach, N. 2007. *Reef Fish Identification: Tropical Pacific*. New World Publications: Jacksonville, USA, 457 pp.

Allen, G.R. and Werner, T.B. 2002. Coral reef fish assessment in the 'coral triangle' of southeastern Asia. *Environmental Biology of Fishes* 65: 209–214.

Gill, A.T., Allen, G.R., and Erdmann, M.V. 2012. Two new dottyback species of the genus *Pseudochromis* from southern Indonesia (Teleosti: Pseudochromidae). *Zootaxa* 3161: 53–60.

Hobbs, J.P.A., Frishch, A.J., Allen, G.R., and van Herwerden, L. In press. Marine hybrid hotspot at Indo-Pacific biogeographic border. *Biology letters*: (doi: 10.1098/rsbl.2008.0561.

Kuiter, R.H. and Tonozuka, T. 2001. *Photo guide to Indonesian reef fishes*. Zoonetics: Seaford, Australia, 893 pp.

Pyle, R.L. and J.E. Randall. 1994. A Review of Hybridization in Marine Angelfishes (Perciformes, Pomacanthidae). *Environmental Biology of Fishes* 41:127–145.

Randall J. E. 1998. Zoogeography of shore fishes of the Indo-Pacific region. *Zoological Studies* 37(4): 227–268.

Smith-Vaniz, W.F. and Allen, G.R. 2011. Three new species of the fangblenny genus *Meiacanthus* from Indonesia, with color photographs and comments on other species. *Zootaxa* 3046: 39–58.

Appendix 3.1. List of the reef fishes of Bali (including Nusa Penida). **New records for Bali are indicated in the left column with a gray box.**

This list includes all species of shallow (to 70 m depth) coral reef fishes known to date from Bali and Nusa Penida. The first twocolumns include species that were previously recorded from Bali (by GRA) and Nusa Pendia (2008 RAP and other surveys). The subsequent site columns refer to the 2011 Bali survey. The phylogenetic sequence of the families appearing in this list follows Eschmeyer (Catalog of Fishes, California Academy of Sciences, 1998) with slight modification (e.g.. placement of Cirrhitidae). Genera and species are arranged alphabetically within each family. The Author name(s) and year of publication have been omitted from each species entry, but this information can be easily accessed on the California Academy of Sciences Catalog of Fishes website: http://www.calacademy.org/research/ichthyology/catalog/fishcatsearch.html.

	Previous Bali	NP Surveys	Site 1	Site 2	Site 3	Site 4	Site 5	Site 7	Site 9	Site 10	Site 11	Site 12	Site 13	Site 14	Site 15	Site 16	Site 17	Site 18	Site 19	Site 20	Site 21	Site 22	Site 23	Site 24	Site 25	Site 26	Site 28	Site 33	Site 29	Site 30	Site 31	Site 32	Present Survey	Previous Surveys	Grand Total	Nusa Penida	East Bali	North Bali	Gilimanuk
Rhincodontidae (1 spp.)																																	0	0	0	0	0	0	0
Rhincodon typus		1																															0	1	1	1	0	0	0
Alopiidae (1 spp.)																																	0	0	0	0	0	0	0
Alopias pelagicus		1																															0	1	1	1	0	0	0
Orectobobidae (1 spp.)																																	0	0	0	0	0	0	0
Orectolobus japonicus		1																															0	1	1	1	0	0	0
Carcharhinidae (2 spp.)																																	0	0	0	0	0	0	0
Carcharhinus amblyrhynchos	1	1																															0	1	1	1	0	0	0
Triaenodon obesus													1													1							1	0	1	0	1	1	0
Dasyatidae (3 spp.)																																	0	0	0	0	0	0	0
Dasyatis kuhlii	1	1	1		1	1						1			1				1	1													1	1	1	1	1	1	0
Taeniura lymma			1	1	1	1		1					1	1	1	1	1	1	1					1		1							1	0	1	0	1	1	0
Taeniura meyeni		1																															0	1	1	1	0	0	0
Myliobatidae (1 spp.)																																	0	0	0	0	0	0	0
Aetobatus narinari		1																															0	1	1	1	0	0	0
Mobulidae (1 spp.)																																	0	0	0	0	0	0	0
Manta birostris		1																															0	1	1	1	0	0	0
Moringuidae (1 spp.)																																	0	0	0	0	0	0	0
Moringa microchir		1																															0	1	1	1	0	0	0
Chlopsidae (1 spp.)																																	0	0	0	0	0	0	0
Kaupichthys diodontus		1																															0	1	1	1	0	0	0
Muraenidae (15 spp.)																																	0	0	0	0	0	0	0
Anarchias seychellensis	1																																0	1	1	0	0	0	0
Echidna nebulosa		1									1										1		1										1	1	1	1	1	1	0

table continued on next page

Appendix 3.1. *continued*

	Previous Bali	NP Surveys	Site 1	Site 2	Site 3	Site 4	Site 5	Site 7	Site 9	Site 10	Site 11	Site 12	Site 13	Site 14	Site 15	Site 16	Site 17	Site 18	Site 19	Site 20	Site 21	Site 22	Site 23	Site 24	Site 25	Site 26	Site 28	Site 33	Site 29	Site 30	Site 31	Site 32	Present Survey	Previous Surveys	Grand Total	Nusa Penida	East Bali	North Bali	Gilimanuk
Gymnothorax angusticauda																						1											1	0	1	0	0	1	0
Gymnothorax chilospilus		1																															0	1	1	1	0	0	0
Gymnothorax fimbriatus		1																	1				1									1	1	1	1	1	0	1	0
Gymnothorax flavimarginatus	1													1																			1	1	1	0	1	0	0
Gymnothorax javanicus	1													1					1								1						1	1	1	0	1	1	0
Gymnothorax melatremus		1																															0	1	1	1	0	0	0
Gymnothorax monochrous																			1														1	0	1	0	0	1	0
Gymnothorax richardsonii?		1																															0	1	1	1	0	0	0
Gymnothorax thrysoideus																							1										1	0	1	0	0	1	0
Gymnothorax zonipectis		1																	1														1	1	1	1	0	1	0
Rhinomuraena quaesita	1			1																1													1	1	1	0	1	1	0
Scuticara tigrina																											1						1	0	1	0	0	1	0
Uropterygius fuscoguttatus		1																															0	1	1	1	0	0	0
Ophichthidae (5 spp.)																																	0	0	0	0	0	0	0
Brachysomophis cirrocheilos																							1										1	0	1	0	0	1	0
Myrichthys maculosus		1																															0	1	1	1	0	0	0
Ophichthus bonaparti																						1	1										1	0	1	0	0	1	0
Pisodonophis cancrivorus																						1											1	0	1	0	0	1	0
Scolecenchelys macroptera																			1														1	0	1	0	0	1	0
Congridae (8 spp.)																																	0	0	0	0	0	0	0
Ariosoma fasciatum												1																					1	0	1	0	1	0	0
Gorgasia barnesi																								1									1	0	1	0	0	1	0
Gorgasia maculata																		1															1	0	1	0	0	1	0
Heteroconger enigmaticus																							1										1	0	1	0	0	1	0
Heteroconger hassi	1	1						1																1									1	1	1	1	1	1	0
Heteroconger mercyae																												1					1	0	1	0	0	1	0
Heteroconger perissodon	1																				1		1										1	1	1	0	0	1	0
Heteroconger polyzona																	1																1	0	1	0	0	1	0

table continued on next page

Appendix 3.1. *continued*

	Previous Bali	NP Surveys	Site 1	Site 2	Site 3	Site 4	Site 5	Site 7	Site 9	Site 10	Site 11	Site 12	Site 13	Site 14	Site 15	Site 16	Site 17	Site 18	Site 19	Site 20	Site 21	Site 22	Site 23	Site 24	Site 25	Site 26	Site 28	Site 33	Site 29	Site 30	Site 31	Site 32	Present Survey	Previous Surveys	Grand Total	Nusa Penida	East Bali	North Bali	Gilimanuk
Clupeidae (1 spp.)																																	0	0	0	0	0	0	0
Spratelloides delicatulus		1																															0	1	1	1	0	0	0
Plotosidae (1 spp.)																																	0	0	0	0	0	0	0
Plotosus lineatus	1	1														1				1				1						1			1	1	1	1	0	1	1
Synodontidae (6 spp.)																																	0	0	0	0	0	0	0
Saurida elongata																							1										1	0	1	0	0	1	0
Saurida gracilis	1	1																															0	1	1	1	0	0	0
Saurida nebulosa																	1						1			1							1	0	1	0	0	1	0
Synodus dermatogenys				1				1							1						1				1	1					1	1	1	0	1	0	1	1	0
Synodus jaculum	1	1												1	1		1					1											1	1	1	1	1	1	0
Synodus variegatus				1																													1	0	1	0	1	0	0
Ophidiidae (1 spp.)																																	0	0	0	0	0	0	0
Ophidiid sp. (deep 70m Menjangan)																										1							1	0	1	0	0	1	0
Bythitidae (2 spp.)																																	0	0	0	0	0	0	0
Diancistrus sp 1 (brown) - live young		1																															0	1	1	1	0	0	0
Nielsenichthys pullus		1																															0	1	1	1	0	0	0
Antennariidae (5 spp.)																																	0	0	0	0	0	0	0
Antennarius coccineus	1	1																															0	1	1	1	0	0	0
Antennarius commersoni	1	1																															0	1	1	1	0	0	0
Antennarius rosaceus																		1															1	0	1	0	0	1	0
Antennarius sp.		1																															0	1	1	1	0	0	0
Antennatus tuberosus													1																				1	0	1	0	1	0	0
Gobiesocidae (3 spp.)																																	0	0	0	0	0	0	0
Diademichthys lineatus																							1										1	0	1	0	0	1	0
Lepadichthys lineatus																										1							1	0	1	0	0	1	0
Lepadichthys species																									1	1							1	0	1	0	0	1	0
Mugilidae (2 spp.)																																	0	0	0	0	0	0	0
Crenimugil crenilabis		1																															0	1	1	1	0	0	0

table continued on next page

Appendix 3.1. *continued*

	Previous Bali	NP Surveys	Site 1	Site 2	Site 3	Site 4	Site 5	Site 7	Site 9	Site 10	Site 11	Site 12	Site 13	Site 14	Site 15	Site 16	Site 17	Site 18	Site 19	Site 20	Site 21	Site 22	Site 23	Site 24	Site 25	Site 26	Site 28	Site 33	Site 29	Site 30	Site 31	Site 32	Present Survey	Previous Surveys	Grand Total	Nusa Penida	East Bali	North Bali	Gilimanuk
Valamugil seheli	1												1			1			1														1	1	1	0	1	1	0
Belonidae (2spp.)																																	0	0	0	0	0	0	0
Tylosurus acus		1																															0	1	1	1	0	0	0
Tylosurus crocodilus	1	1						1						1			1									1							1	1	1	1	1	1	0
Hemiramphidae (1 spp.)																																	0	0	0	0	0	0	0
Hyporhamphus dussumieri		1																															0	1	1	1	0	0	0
Anomalopidae (2 spp.)																																	0	0	0	0	0	0	0
Anomalops katoptron	1																																0	1	1	0	0	0	0
Photoblepharon palpebratum	1																																0	1	1	0	0	0	0
Holocentridae (19 spp.)																																	0	0	0	0	0	0	0
Myripristis berndti	1	1			1	1	1	1		1	1	1	1	1				1	1														1	1	1	1	1	1	0
Myripristis botche	1	1													1																		1	1	1	1	0	1	0
Myripristes chryseres		1												1																			1	1	1	1	1	0	0
Myripristis hexagona																1				1	1			1									1	0	1	0	0	1	0
Myripristis kuntee	1	1	1	1		1		1		1			1		1	1			1														1	1	1	1	1	1	0
Myripristis murdjan	1	1	1	1				1		1	1							1	1		1												1	1	1	1	1	1	0
Myripristis pralinia																1		1															1	0	1	0	0	1	0
Myripristis violacea	1	1																															0	1	1	1	0	0	0
Myripristis vittata	1	1														1		1															1	1	1	1	0	1	0
Neoniphon aurolineatus		1																															0	1	1	1	0	0	0
Neoniphon sammara		1			1	1	1				1		1	1	1			1															1	1	1	1	1	1	0
Plectrypops lima		1																															0	1	1	1	0	0	0
Sargocentron caudimaculatum	1	1				1	1	1		1	1	1	1	1	1	1	1	1	1		1			1	1	1	1						1	1	1	1	1	1	0
Sargocentron diadema	1	1									1																						1	1	1	1	1	0	0
Sargocentron ittodai	1	1																															0	1	1	1	0	0	0
Sargocentron melanospilos											1																						1	0	1	0	1	0	0
Sargocentron microstoma								1																									1	0	1	0	1	0	0
Sargocentron praslin	1																																0	1	1	0	0	0	0

table continued on next page

Appendix 3.1. *continued*

	Previous Bali	NP Surveys	Site 1	Site 2	Site 3	Site 4	Site 5	Site 7	Site 9	Site 10	Site 11	Site 12	Site 13	Site 14	Site 15	Site 16	Site 17	Site 18	Site 19	Site 20	Site 21	Site 22	Site 23	Site 24	Site 25	Site 26	Site 28	Site 33	Site 29	Site 30	Site 31	Site 32	Present Survey	Previous Surveys	Grand Total	Nusa Penida	East Bali	North Bali	Gilimanuk
Sargocentron rubrum				1								1																				1	1	0	1	0	1	0	0
Pegasidae (2 spp.)																																	0	0	0	0	0	0	0
Eurypegasus draconis																													1				1	0	1	0	0	0	1
Pegasus volitans																													1				1	0	1	0	0	0	1
Aulostomidae (1 spp.)																																	0	0	0	0	0	0	0
Aulostomus chinensis	1	1			1	1	1	1	1	1	1	1	1	1				1	1					1		1	1						1	1	1	1	1	1	0
Fistulariidae (1 spp.)																		1	1														1	0	1	0	0	1	0
Fistularia commersonii	1	1					1		1		1		1											1		1	1						1	1	1	1	1	1	0
Centriscidae (3 spp.)																																	0	0	0	0	0	0	0
Aeoliscus strigatus																										1			1	1			1	0	1	0	0	1	1
Centriscus cristatus																														1			1	0	1	0	0	0	1
Centriscus scutatus				1	1														1														1	0	1	0	1	1	0
Syngnathidae (6 spp.)																																	0	0	0	0	0	0	0
Corythoichthys haematopterus																									1								1	0	1	0	0	1	0
Doryrhamphus melanopleura																															1		1	0	1	0	0	0	0
Dunckerocampus dactyliophorus	1				1																								1				1	1	1	0	1	0	1
Hippocampus histrix																			1														1	0	1	0	0	1	0
Hippocampus kuda																													1				1	0	1	0	0	0	1
Trachyrhamphus bicoarctatus																													1				1	0	1	0	0	0	1
Scorpaenidae (21 spp.)																																	0	0	0	0	0	0	0
Dendrochirus brachypterus																							1						1				1	0	1	0	0	1	1
Dendrochirus zebra		1					1			1						1							1						1			1	1	1	1	1	1	1	1
Parascorpaena picta																	1						1										1	0	1	0	0	1	0
Pterois antennata	1	1											1			1		1	1		1				1	1							1	1	1	1	1	1	0
Pterois mombasae		1																	1														1	1	1	1	0	1	0
Pterois radiata		1								1																							1	1	1	1	1	0	0
Pterois russellii	1										1											1	1						1				1	1	1	0	1	1	1
Pterois volitans	1								1													1				1	1		1				1	1	1	0	1	1	1

table continued on next page

Appendix 3.1. *continued*

	Previous Bali	NP Surveys	Site 1	Site 2	Site 3	Site 4	Site 5	Site 7	Site 9	Site 10	Site 11	Site 12	Site 13	Site 14	Site 15	Site 16	Site 17	Site 18	Site 19	Site 20	Site 21	Site 22	Site 23	Site 24	Site 25	Site 26	Site 28	Site 33	Site 29	Site 30	Site 31	Site 32	Present Survey	Previous Surveys	Grand Total	Nusa Penida	East Bali	North Bali	Gilimanuk
Scorpaenodes evides		1																															0	1	1	1	0	0	0
Scorpaenodes guamensis		1																															0	1	1	1	0	0	0
Scorpaenodes hirsutus		1																															0	1	1	1	0	0	0
Scorpaenodes kelloggi								1																									1	0	1	0	1	0	0
Scorpaenodes parvipinnis		1						1																									1	1	1	1	1	0	0
Scorpaenodes varipinnis	1						1																										1	1	1	0	1	0	0
Scorpaenopsis diabolus								1																									1	0	1	0	1	0	0
Scorpaenopsis macrochir	1																																0	1	1	0	0	0	0
Scorpaenopsis neglecta		1																															0	1	1	1	0	0	0
Scorpaenopsis oxycephala	1	1			1		1			1			1																				1	1	1	1	1	0	0
Scorpaenopsis papuensis					1					1															1								1	0	1	0	1	1	0
Scorpaenopsis possi								1																									1	0	1	0	1	0	0
Taenianotus triacanthus		1		1																													1	1	1	1	1	0	0
Synanceiidae (1 spp.)																																	0	0	0	0	0	0	0
Inimicus didactylus																						1									1		1	0	1	0	0	1	0
Tetrarogidae (2 spp.)																																	0	0	0	0	0	0	0
Ablabys macracanthus																						1											1	0	1	0	0	1	0
Ablabys taenianotus				1																													1	0	1	0	1	0	0
Platycephalidae (6 spp.)																																	0	0	0	0	0	0	0
Cociella punctata		1																															0	1	1	1	0	0	0
Cymbacephalus beauforti																								1									1	0	1	0	0	1	0
Eurycephalus arenicola		1																															0	1	1	1	0	0	0
Onigocia pedimacula																									1								1	0	1	0	0	1	0
Onigocia sp. collected		1																															0	1	1	1	0	0	0
Thysanophrys chiltonae		1																															0	1	1	1	0	0	0
Caracanthidae (1 spp.)																																	0	0	0	0	0	0	0
Caracanthus unipinna	1																																0	1	1	0	0	0	0
Dactylopteridae (1 spp.)																																	0	0	0	0	0	0	0

table continued on next page

Appendix 3.1. *continued*

	Previous Bali	NP Surveys	Site 1	Site 2	Site 3	Site 4	Site 5	Site 7	Site 9	Site 10	Site 11	Site 12	Site 13	Site 14	Site 15	Site 16	Site 17	Site 18	Site 19	Site 20	Site 21	Site 22	Site 23	Site 24	Site 25	Site 26	Site 28	Site 33	Site 29	Site 30	Site 31	Site 32	Present Survey	Previous Surveys	Grand Total	Nusa Penida	East Bali	North Bali	Gilimanuk
Dactyloptena orientalis																	1																1	0	1	0	0	1	0
Centrogeniidae (1 spp.)																																	0	0	0	0	0	0	0
Centrogenys vaigiensis	1																																0	1	1	0	0	0	0
Serranidae (54 spp.)																																	0	0	0	0	0	0	0
Aethaloperca rogaa	1	1					1								1	1	1	1		1	1			1		1	1						1	1	1	1	1	1	0
Anyperodon leucogrammicus												1	1	1	1	1	1	1							1	1	1				1		1	0	1	0	1	1	0
Belonoperca chabanaudi																		1															1	0	1	0	0	1	0
Cephalopholis argus	1	1									1		1											1		1	1						1	1	1	1	1	1	0
Cephalopholis boenak		1																															0	1	1	1	0	0	0
Cephalopholis cyanostigma									1				1		1					1	1			1	1	1	1				1	1	1	0	1	0	1	1	0
Cephalopholis leopardus		1													1		1	1							1								1	1	1	1	0	1	0
Cephalopholis microprion	1																								1	1			1				1	1	1	0	0	1	1
Cephalopholis miniata		1								1		1	1	1	1	1	1	1	1	1	1					1							1	1	1	1	1	1	0
Cephalopholis sexmaculata		1														1		1		1				1			1						1	1	1	1	0	1	0
Cephalopholis sonnerati																	1		1														1	0	1	0	0	1	0
Cephalopholis spiloparaea														1																			1	0	1	0	1	0	0
Cephalopholis urodeta	1	1				1	1							1	1	1	1	1	1					1	1								1	1	1	1	1	1	0
Epinephelus areolatus	1	1							1							1						1	1	1									1	1	1	1	1	1	0
Epinephelus bontoides																1			1														1	0	1	0	0	1	0
Epinephelus caeruleopunctatus									1																								1	0	1	0	1	0	0
Epinephelus coioides				1																			1						1				1	0	1	0	1	1	1
Epinephelus fasciatus		1						1							1		1		1		1			1	1	1					1	1	1	1	1	1	1	1	0
Epinephelus fuscoguttatus		1																															0	1	1	1	0	0	0
Epinephelus lanceolatus		1																															0	1	1	1	0	0	0
Epinephelus maculatus				1					1										1													1	1	0	1	0	1	1	0
Epinephelus melanostigma		1																															0	1	1	1	0	0	0
Epinephelus merra	1	1	1	1	1				1									1									1						1	1	1	1	1	1	0
Epinephelus ongus																					1												1	0	1	0	0	1	0

table continued on next page

Appendix 3.1. *continued*

	Previous Bali	NP Surveys	Site 1	Site 2	Site 3	Site 4	Site 5	Site 7	Site 9	Site 10	Site 11	Site 12	Site 13	Site 14	Site 15	Site 16	Site 17	Site 18	Site 19	Site 20	Site 21	Site 22	Site 23	Site 24	Site 25	Site 26	Site 28	Site 33	Site 29	Site 30	Site 31	Site 32	Present Survey	Previous Surveys	Grand Total	Nusa Penida	East Bali	North Bali	Gilimanuk
Epinephelus quoyanus																																1	1	0	1	0	0	0	0
Epinephelus undulosus																						1											1	0	1	0	0	1	0
Grammistes sexlineatus	1	1									1			1			1						1										1	1	1	1	1	1	0
Luzonichthys waitei																				1													1	0	1	0	0	1	0
Plectranthias inermis																									1								1	0	1	0	0	1	0
Plectranthias longimanus	1	1									1		1					1	1														1	1	1	1	1	1	0
Plectropomus laevis																										1							1	0	1	0	0	1	0
Plectropomus leopardus																										1							1	0	1	0	0	1	0
Plectropomus maculatus					1																											1	1	0	1	0	1	0	0
Pogonoperca punctata		1																															0	1	1	1	0	0	0
Pseudanthias bicolor		1															1									1							1	1	1	1	0	1	0
Pseudanthias bimaculatus	1																																0	1	1	0	0	0	0
Pseudanthias charleneae	1	1													1												1						1	1	1	1	0	1	0
Pseudanthias dispar		1						1			1		1	1				1			1												1	1	1	1	1	1	0
Pseudanthias fasciatus		1												1	1	1	1										1						1	1	1	1	1	1	0
Pseudanthias huchtii	1	1								1	1		1	1	1	1	1	1	1	1	1			1	1	1	1				1	1	1	1	1	1	1	1	0
Pseudanthias hutomoi																				1						1	1						1	0	1	0	0	1	0
Pseudanthias hypselosoma		1						1	1				1	1	1	1	1			1				1		1					1		1	1	1	1	1	1	0
Pseudanthias lori		1																									1						1	1	1	1	0	1	0
Pseudanthias luzonensis	1	1						1		1					1																		1	1	1	1	1	1	0
Pseudanthias parvirostris		1																															0	1	1	1	0	0	0
Pseudanthias pleurotaenia	1	1			1					1		1	1	1	1	1	1	1	1	1				1		1	1						1	1	1	1	1	1	0
Pseudanthias randalli		1																															0	1	1	1	0	0	0
Pseudanthias squamipinnis	1	1					1	1	1	1	1	1	1	1				1	1					1	1	1	1				1	1	1	1	1	1	1	1	0
Pseudanthias tuka		1						1					1			1		1			1									1			1	1	1	1	1	1	1
Pseudogramma polyacanthus		1									1								1														1	1	1	1	1	1	0
Pseudogramma sp. 70 m (photo)		1																															0	1	1	1	0	0	0
Serranocirrhitus latus		1																															0	1	1	1	0	0	0

table continued on next page

Appendix 3.1. *continued*

	Previous Bali	NP Surveys	Site 1	Site 2	Site 3	Site 4	Site 5	Site 7	Site 9	Site 10	Site 11	Site 12	Site 13	Site 14	Site 15	Site 16	Site 17	Site 18	Site 19	Site 20	Site 21	Site 22	Site 23	Site 24	Site 25	Site 26	Site 28	Site 33	Site 29	Site 30	Site 31	Site 32	Present Survey	Previous Surveys	Grand Total	Nusa Penida	East Bali	North Bali	Gilimanuk
Variola albimarginata	1	1						1		1	1	1	1	1	1	1		1	1					1		1							1	1	1	1	1	1	0
Variola louti		1						1				1	1		1	1	1	1	1														1	1	1	1	1	1	0
Cirrhitidae (7 spp.)																																	0	0	0	0	0	0	0
Cirrhitichthys aprinus	1	1																				1							1		1	1	1	1	1	1	0	1	1
Cirrhitichthys falco		1		1													1																1	1	1	1	1	1	0
Cirrhitichthys oxycephalus		1					1			1	1	1	1	1					1	1													1	1	1	1	1	1	0
Cirrhitus pinnulatus	1	1								1				1		1																	1	1	1	1	1	1	0
Cyprinocirrhites polyactis	1	1								1				1	1	1	1							1			1						1	1	1	1	1	1	0
Paracirrhites arcatus		1									1				1	1	1	1	1														1	1	1	1	1	1	0
Paracirrhites forsteri	1	1	1	1	1	1	1	1		1	1	1	1	1	1	1	1	1	1					1		1	1						1	1	1	1	1	1	0
Pseudochromidae (19 spp.)																																	0	0	0	0	0	0	0
Congrogadus subducens		1																															0	1	1	1	0	0	0
Haliophis aethiopus	1	1																															0	1	1	1	0	0	0
Labracinus cyclophthalmus	1	1			1		1																							1			1	1	1	1	1	0	1
Lubbockichthys multisquamatus																											1						1	0	1	0	0	1	0
Manonichthys sp. 1 (cf. alleni)																									1		1						1	0	1	0	0	1	0
Pictichromis paccagnellae												1				1		1			1			1	1	1	1						1	0	1	0	1	1	0
Pseudochromis andamanensis			1				1				1																						1	0	1	0	1	0	0
Pseudochromis arulenteus	1	1								1	1																						1	1	1	1	1	0	0
Pseudochromis fuscus		1																												1			1	1	1	1	0	0	1
Pseudochromis litus		1																							1								1	1	1	1	0	1	0
Pseudochromis marshallensis																								1	1		1						1	0	1	0	0	1	0
Pseudochromis oligochrysus		1																								1							1	1	1	1	0	1	0
Pseudochromis perspicillatus																																1	1	0	1	0	0	0	0
Pseudochromis ransonetti																															1		1	0	1	0	0	0	0
Pseudochromis rutilus	1	1																															0	1	1	1	0	0	0
Pseudochromis steenei																			1														1	0	1	0	0	1	0
Pseudoplesiops annae																									1								1	0	1	0	0	1	0

table continued on next page

Appendix 3.1. *continued*

	Previous Bali	NP Surveys	Site 1	Site 2	Site 3	Site 4	Site 5	Site 7	Site 9	Site 10	Site 11	Site 12	Site 13	Site 14	Site 15	Site 16	Site 17	Site 18	Site 19	Site 20	Site 21	Site 22	Site 23	Site 24	Site 25	Site 26	Site 28	Site 33	Site 29	Site 30	Site 31	Site 32	Present Survey	Previous Surveys	Grand Total	Nusa Penida	East Bali	North Bali	Gilimanuk
Pseudoplesiops collare																								1									1	0	1	0	0	1	0
Pseudoplesiops immaculatus		1																															0	1	1	1	0	0	0
Plesiopidae (4 spp.)																																	0	0	0	0	0	0	0
Belonopterygium fasciolatum		1																															0	1	1	1	0	0	0
Calloplesiops altivelis																											1						1	0	1	0	0	1	0
Plesiops coeruleolineatus		1					1																										1	1	1	1	1	0	0
Steeneichthys nativitatis						1																											1	0	1	0	1	0	0
Opistognathidae (5 spp.)																																	0	0	0	0	0	0	0
Opistognathus sp. 1 "hyalinus"																									1								1	0	1	0	0	1	0
Opistognathus randalli																	1							1		1							1	0	1	0	0	1	0
Opistognathus solorensis							1																										1	0	1	0	1	0	0
Opistognathus variabilis			1																														1	0	1	0	1	0	0
Opistognathus sp. 2 "vicinus"																									1								1	0	1	0	0	1	0
Priacanthidae (3 spp.)																																	0	0	0	0	0	0	0
Priacanthus blochii			1	1			1																										1	0	1	0	1	0	0
Priacanthus hamrur				1						1																							1	0	1	0	1	0	0
Priacanthus sagittarius																					1												1	0	1	0	0	1	0
Apogonidae (59 spp.)																																	0	0	0	0	0	0	0
Apogon angustatus											1		1		1				1														1	0	1	0	1	1	0
Apogon apogonides	1	1														1																	1	1	1	1	0	1	0
Apogon aureus	1	1		1	1											1				1	1	1										1	1	1	1	1	1	1	0
Apogon bryx									1													1	1										1	0	1	0	1	1	0
Apogon ceramensis																													1				1	0	1	0	0	0	1
Apogon chrysopomus				1	1																												1	0	1	0	1	0	0
Apogon chrysotaenia	1	1																	1												1	1	1	1	1	1	0	1	0
Apogon compressus				1																	1									1			1	0	1	0	1	1	1
Apogon crassiceps	1	1																															0	1	1	1	0	0	0
Apogon cyanosoma																								1					1				1	0	1	0	0	1	1

table continued on next page

Appendix 3.1. *continued*

	Previous Bali	NP Surveys	Site 1	Site 2	Site 3	Site 4	Site 5	Site 7	Site 9	Site 10	Site 11	Site 12	Site 13	Site 14	Site 15	Site 16	Site 17	Site 18	Site 19	Site 20	Site 21	Site 22	Site 23	Site 24	Site 25	Site 26	Site 28	Site 33	Site 29	Site 30	Site 31	Site 32	Present Survey	Previous Surveys	Grand Total	Nusa Penida	East Bali	North Bali	Gilimanuk
Apogon dispar	1																										1						1	1	1	0	0	1	0
Apogon evermanni	1	1																															0	1	1	1	0	0	0
Apogon exostigma	1																																0	1	1	0	0	0	0
Apogon fleurieu																				1				1									1	0	1	0	0	1	0
Apogon fraenatus	1			1													1			1	1												1	1	1	0	1	1	0
Apogon guamensis																			1		1								1	1			1	0	1	0	0	1	1
Apogon hartzfeldii																													1	1			1	0	1	0	0	0	1
Apogon hoevenii																													1	1			1	0	1	0	0	0	1
Apogon kallopterus	1	1		1											1		1		1	1	1												1	1	1	1	1	1	0
Apogon leptacanthus																													1	1			1	0	1	0	0	0	1
Apogon lineomaculus																				1													1	0	1	0	0	1	0
Apogon monospilus				1																		1	1										1	0	1	0	1	1	0
Apogon moluccensis	1																					1	1										1	1	1	0	0	1	0
Apogon multilineatus		1																															0	1	1	1	0	0	0
Apogon nigrofasciatus	1	1													1			1	1	1	1				1	1	1				1	1	1	1	1	1	0	1	0
Apogon novemfasciatus		1																															0	1	1	1	0	0	0
Apogon parvulus																1								1						1			1	0	1	0	0	1	1
Apogon schlegeli		1																															0	1	1	1	0	0	0
Apgon seminigracaudus		1						1		1									1						1								1	1	1	1	1	1	0
Apogon semiornatus		1																	1														1	1	1	1	0	1	0
Apogon taeniophorus						1											1		1	1	1												1	0	1	0	1	1	0
Apogon thermalis																													1	1			1	0	1	0	0	0	1
Apogon timorensis		1																															0	1	1	1	0	0	0
Apogon trimaculatus					1																												1	0	1	0	1	0	0
Apogon viria																													1	1			1	0	1	0	0	0	1
Apogon wassinki	1	1															1			1				1							1		1	1	1	1	0	1	0
Apogonichthys perdix		1																															0	1	1	1	0	0	0
Archamia biguttata				1																	1			1									1	0	1	0	1	1	0

table continued on next page

Appendix 3.1. *continued*

	Previous Bali	NP Surveys	Site 1	Site 2	Site 3	Site 4	Site 5	Site 7	Site 9	Site 10	Site 11	Site 12	Site 13	Site 14	Site 15	Site 16	Site 17	Site 18	Site 19	Site 20	Site 21	Site 22	Site 23	Site 24	Site 25	Site 26	Site 28	Site 33	Site 29	Site 30	Site 31	Site 32	Present Survey	Previous Surveys	Grand Total	Nusa Penida	East Bali	North Bali	Gilimanuk
Archamia fucata				1																1	1												1	0	1	0	1	1	0
Archamia macroptera				1																									1	1			1	0	1	0	1	0	1
Archamia melasma																					1												1	0	1	0	0	1	0
Cheilodipterus artus									1											1									1	1			1	0	1	0	1	1	1
Cheilodipterus macrodon		1					1		1											1				1	1		1		1	1			1	1	1	1	1	1	1
Cheilodipterus quinquelineatus	1	1		1	1				1						1														1	1			1	1	1	1	1	1	1
Coranthus polyacanthus		1																															0	1	1	1	0	0	0
Foa fo									1											1													1	0	1	0	1	1	0
Fowleria marmorata		1																							1								1	1	1	1	0	1	0
Fowleria vaiulae														1																			1	0	1	0	1	0	0
Fowleria variegata		1																															0	1	1	1	0	0	0
Neamia notula																									1								1	0	1	0	0	1	0
Neamia octospina		1																															0	1	1	1	0	0	0
Pseudamia gelatinosa		1																															0	1	1	1	0	0	0
Pseudamiops gracilicauda																			1														1	0	1	0	0	1	0
Pterapogon kauderni	1																												1				1	1	1	0	0	0	1
Rhabdamia cypselurus	1																																0	1	1	0	0	0	0
Rhabdamia gracilis	1	1												1										1									1	1	1	1	1	1	0
Siphamia sp. 1 (cf argentea 70 m Menjangan)																										1							1	0	1	0	0	1	0
Siphamia tubifer																							1										1	0	1	0	0	1	0
Sphaeramia nematoptera																														1			1	0	1	0	0	0	1
Malacanthidae (6 spp.)																																	0	0	0	0	0	0	0
Hoplolatilus chlupatyi								1																									1	0	1	0	1	0	0
Hoplolatilus cuniculus		1								1	1															1	1						1	1	1	1	1	1	0
Hoplolatilus randalli		1								1																							1	1	1	1	1	0	0
Hoplolatilus starcki	1	1												1				1															1	1	1	1	1	1	0
Malacanthus brevirostris		1						1		1		1	1	1	1	1	1	1						1		1							1	1	1	1	1	1	0

table continued on next page

Appendix 3.1. *continued*

	Previous Bali	NP Surveys	Site 1	Site 2	Site 3	Site 4	Site 5	Site 7	Site 9	Site 10	Site 11	Site 12	Site 13	Site 14	Site 15	Site 16	Site 17	Site 18	Site 19	Site 20	Site 21	Site 22	Site 23	Site 24	Site 25	Site 26	Site 28	Site 33	Site 29	Site 30	Site 31	Site 32	Present Survey	Previous Surveys	Grand Total	Nusa Penida	East Bali	North Bali	Gilimanuk
Malacanthus latovittatus	1	1						1							1	1	1	1															1	1	1	1	1	1	0
Echeneidae (1 spp.)																																	0	0	0	0	0	0	0
Echeneis naucrates	1	1																				1											1	1	1	1	0	1	0
Carangidae (10 spp.)																																	0	0	0	0	0	0	0
Carangoides bajad															1	1	1									1	1						1	0	1	0	0	1	0
Carangoides ferdau				1															1														1	0	1	0	1	1	0
Carangoides fulvoguttatus	1																																0	1	1	0	0	0	0
Carangoides oblongus				1																													1	0	1	0	1	0	0
Carangoides plagiotaenia		1						1					1					1								1	1						1	1	1	1	1	1	0
Caranx ignobilis	1	1																	1														1	1	1	1	0	1	0
Caranx melampygus	1	1		1				1		1		1	1	1	1	1	1	1	1						1	1	1						1	1	1	1	1	1	0
Caranx sexfasciatus																		1															1	0	1	0	0	1	0
Elagatis bipinnulata										1																							1	0	1	0	1	0	0
Scomberoides lysan															1				1							1							1	0	1	0	0	1	0
Lutjanidae (22 spp.)																																	0	0	0	0	0	0	0
Aprion virescens		1																															0	1	1	1	0	0	0
Lutjanus argentimaculatus				1																	1												1	0	1	0	1	1	0
Lutjanus biguttatus								1																									1	0	1	0	1	0	0
Lutjanus bohar	1	1										1	1	1	1	1	1	1		1				1	1	1	1						1	1	1	1	1	1	0
Lutjanus decussatus	1	1						1		1			1	1	1	1	1	1	1	1	1			1	1	1	1				1	1	1	1	1	1	1	1	0
Lutjanus ehrenbergii														1																			1	0	1	0	1	0	0
Lutjanus fulviflamma	1	1	1	1	1										1	1	1	1	1							1	1						1	1	1	1	1	1	0
Lutjanus fulvus	1	1		1	1									1	1	1	1	1	1		1			1		1							1	1	1	1	1	1	0
Lutjanus gibbus	1	1				1									1	1		1						1		1			1				1	1	1	1	1	1	1
Lutjanus kasmira	1	1						1		1				1	1	1										1							1	1	1	1	1	1	0
Lutjanus lutjanus														1			1					1										1	1	0	1	0	1	1	0
Lutjanus madras																1	1	1															1	0	1	0	0	1	0
Lutjanus malabaricus																							1										1	0	1	0	0	1	0

table continued on next page

Appendix 3.1. *continued*

	Previous Bali	NP Surveys	Site 1	Site 2	Site 3	Site 4	Site 5	Site 7	Site 9	Site 10	Site 11	Site 12	Site 13	Site 14	Site 15	Site 16	Site 17	Site 18	Site 19	Site 20	Site 21	Site 22	Site 23	Site 24	Site 25	Site 26	Site 28	Site 33	Site 29	Site 30	Site 31	Site 32	Present Survey	Previous Surveys	Grand Total	Nusa Penida	East Bali	North Bali	Gilimanuk
Lutjanus monostigma		1								1																1							1	1	1	1	1	1	0
Lutjanus quinquelineatus					1														1			1	1			1			1				1	0	1	0	1	1	1
Lutjanus rivulatus	1	1		1									1		1	1	1	1															1	1	1	1	1	1	0
Lutjanus rufolineatus																			1										1				1	0	1	0	0	1	1
Lutjanus sebae																						1	1										1	0	1	0	0	1	0
Macolor macularis	1	1								1	1		1	1	1	1	1	1	1	1				1		1	1						1	1	1	1	1	1	0
Macolor niger	1	1																						1		1							1	1	1	1	0	1	0
Paracaesio sordida																									1								1	0	1	0	0	1	0
Paracaesio xanthura		1			1					1																							1	1	1	1	1	0	0
Caesionidae (14 spp.)																																	0	0	0	0	0	0	0
Caesio caerulaurea	1	1	1			1		1		1	1		1	1	1	1	1	1						1		1	1		1			1	1	1	1	1	1	1	1
Caesio cuning				1						1		1		1	1	1	1		1					1		1	1				1	1	1	0	1	0	1	1	0
Caesio lunaris	1	1						1		1	1		1	1	1	1	1	1								1	1						1	1	1	1	1	1	0
Caesio teres	1	1								1	1			1	1	1	1	1								1							1	1	1	1	1	1	0
Caesio varilineata								1																									1	0	1	0	1	0	0
Caesio xanthonota	1	1						1			1							1															1	1	1	1	1	1	0
Pterocaesio chrysozona																															1	1	1	0	1	0	0	0	0
Pterocaesio diagramma	1	1		1	1					1															1								1	1	1	1	1	1	0
Pterocaesio marri																										1							1	0	1	0	0	1	0
Pterocaesio pisang																				1							1			1	1		1	0	1	0	0	1	1
Pterocaesio randalli										1								1							1	1							1	0	1	0	1	1	0
Pterocaesio tessellata			1							1																1	1						1	0	1	0	1	1	0
Pterocaesio tile		1				1		1					1				1	1									1						1	1	1	1	1	1	0
Pterocaesio trilineata																								1						1			1	0	1	0	0	1	1
Symphysanodontidae (1 spp.)																																	0	0	0	0	0	0	0
Symphysanodon cf katayamai		1																															0	1	1	1	0	0	0
Gerreidae (1 spp.)																																	0	0	0	0	0	0	0
Gerres oyena																											1		1				1	0	1	0	0	1	1

table continued on next page

Appendix 3.1. *continued*

	Previous Bali	NP Surveys	Site 1	Site 2	Site 3	Site 4	Site 5	Site 7	Site 9	Site 10	Site 11	Site 12	Site 13	Site 14	Site 15	Site 16	Site 17	Site 18	Site 19	Site 20	Site 21	Site 22	Site 23	Site 24	Site 25	Site 26	Site 28	Site 33	Site 29	Site 30	Site 31	Site 32	Present Survey	Previous Surveys	Grand Total	Nusa Penida	East Bali	North Bali	Gilimanuk
Haemulidae (8 spp.)																																	0	0	0	0	0	0	0
Diagramma pictum																	1												1	1			1	0	1	0	0	1	1
Plectorhinchus chaetodontoides		1		1							1							1								1			1				1	1	1	1	1	1	1
Plectorhinchus chrysotaenia		1																															0	1	1	1	0	0	0
Plectorhinchus flavomaculatus						1																				1							1	0	1	0	1	1	0
Plectorhinchus lessonii	1	1	1	1	1	1	1			1	1		1	1				1							1				1				1	1	1	1	1	1	1
Plectorhinchus lineatus		1		1	1	1				1					1	1	1							1							1	1	1	1	1	1	1	1	0
Plectorhinchus polytaenia	1	1						1	1		1		1			1			1		1					1						1	1	1	1	1	1	1	0
Plectorhinchus vittatus	1	1	1	1	1	1	1			1	1	1	1	1	1	1	1	1	1													1	1	1	1	1	1	1	0
Lethrinidae (10 spp.)																																	0	0	0	0	0	0	0
Gnathodentex aurolineatus		1		1	1					1																							1	1	1	1	1	0	0
Gymnocranius griseus																			1				1										1	0	1	0	0	1	0
Gymnocranius sp.						1				1																							1	0	1	0	1	0	0
Lethrinus amboinensis	1	1																															0	1	1	1	0	0	0
Lethrinus harak	1	1			1											1	1			1			1										1	1	1	1	1	1	0
Lethrinus microdon	1														1																		1	1	1	0	0	1	0
Lethrinus olivaceus		1		1	1			1		1					1			1	1					1								1	1	1	1	1	1	1	0
Lethrinus ornatus		1				1							1		1	1	1	1						1		1							1	1	1	1	1	1	0
Monotaxis grandoculis	1	1		1	1			1			1		1	1	1	1	1	1	1	1				1	1	1	1						1	1	1	1	1	1	0
Monotaxis heterodon													1		1		1	1						1									1	0	1	0	1	1	0
Nemipteridae (13 spp.)																																	0	0	0	0	0	0	0
Pentapodus aureofasciatus	1																							1									1	1	1	0	0	1	0
Pentapodus nagasakiensis ?								1			1																						1	0	1	0	1	0	0
Pentapodus trivittatus		1		1																					1		1		1				1	1	1	1	1	1	1
Scolopsis affinis	1	1	1	1	1			1	1						1	1	1	1	1	1	1	1	1	1		1						1	1	1	1	1	1	1	0
Scolopsis auratus							1					1	1	1	1	1	1	1	1														1	0	1	0	1	1	0
Scolopsis bilineatus	1	1	1	1	1	1	1	1	1	1	1	1	1	1	1	1	1	1	1	1	1		1	1	1	1	1		1	1	1	1	1	1	1	1	1	1	1
Scolopsis ciliatus				1	1				1									1	1						1				1			1	1	0	1	0	1	1	1

table continued on next page

Appendix 3.1. *continued*

	Previous Bali	NP Surveys	Site 1	Site 2	Site 3	Site 4	Site 5	Site 7	Site 9	Site 10	Site 11	Site 12	Site 13	Site 14	Site 15	Site 16	Site 17	Site 18	Site 19	Site 20	Site 21	Site 22	Site 23	Site 24	Site 25	Site 26	Site 28	Site 33	Site 29	Site 30	Site 31	Site 32	Present Survey	Previous Surveys	Grand Total	Nusa Penida	East Bali	North Bali	Gilimanuk
Scolopsis lineatus	1	1		1															1						1	1	1						1	1	1	1	1	1	0
Scolopsis margaritifer																									1					1			1	0	1	0	0	1	1
Scolopsis monogramma	1				1		1								1				1														1	1	1	0	1	1	0
Scolopsis torquata																															1	1	1	0	1	0	0	0	0
Scolopsis trilineatus	1		1																														1	1	1	0	1	0	0
Scolopsis xenochrous	1						1	1		1		1	1	1											1								1	1	1	0	1	1	0
Mullidae (15 spp.)																																	0	0	0	0	0	0	0
Mulloidichthys flavolineatus	1	1	1	1	1								1	1	1	1	1	1						1		1							1	1	1	1	1	1	0
Mulloidichthys vanicolensis	1	1		1		1	1			1		1	1	1	1	1	1	1															1	1	1	1	1	1	0
Parupeneus barberinoides																																1	1	0	1	0	0	0	0
Parupeneus barberinus	1	1	1	1	1	1	1	1	1	1		1	1		1	1	1			1	1			1	1	1	1		1	1	1	1	1	1	1	1	1	1	1
Parupeneus crassilabris		1												1			1							1	1	1	1						1	1	1	1	1	1	0
Parupeneus cyclostomus				1	1		1	1	1	1	1	1	1	1																			1	0	1	0	1	0	0
Parupeneus heptacanthus	1	1	1	1	1		1										1		1	1		1	1										1	1	1	1	1	1	0
Parupeneus indicus	1	1		1	1	1																											1	1	1	1	1	0	0
Parupeneus macronemus	1	1	1	1	1	1	1		1			1	1		1	1	1	1	1					1	1		1						1	1	1	1	1	1	0
Parupeneus multifasciatus	1	1		1	1	1	1	1	1	1	1	1	1		1	1	1	1	1	1	1					1	1				1	1	1	1	1	1	1	1	0
Parupeneus pleurostigma																								1								1	1	0	1	0	0	1	0
Parupeneus spilurus		1						1			1		1																				1	1	1	1	1	0	0
Parupeneus trifasciatus	1	1	1		1		1				1	1																					1	1	1	1	1	0	0
Upeneus sundaicus																						1	1										1	0	1	0	0	1	0
Upeneus tragula					1																								1			1	1	0	1	0	1	0	1
Pempheridae (4 spp.)																																	0	0	0	0	0	0	0
Parapriacanthus ransonneti	1	1		1																													1	1	1	1	1	0	0
Pempheris oualensis	1			1	1																												1	1	1	0	1	0	0
Pempheris schwenkii					1																												1	0	1	0	1	0	0
Pempheris vanicolensis	1	1	1	1	1	1	1								1			1			1										1	1	1	1	1	1	1	1	0
Kyphosidae (2 spp.)																																	0	0	0	0	0	0	0

table continued on next page

Appendix 3.1. *continued*

	Previous Bali	NP Surveys	Site 1	Site 2	Site 3	Site 4	Site 5	Site 7	Site 9	Site 10	Site 11	Site 12	Site 13	Site 14	Site 15	Site 16	Site 17	Site 18	Site 19	Site 20	Site 21	Site 22	Site 23	Site 24	Site 25	Site 26	Site 28	Site 33	Site 29	Site 30	Site 31	Site 32	Present Survey	Previous Surveys	Grand Total	Nusa Penida	East Bali	North Bali	Gilimanuk
Kyphosus cinerascens		1								1						1		1						1									1	1	1	1	1	1	0
Kyphosus vaigensis	1	1		1	1			1				1		1	1	1	1	1								1	1						1	1	1	1	1	1	0
Monodactylidae (1 spp.)																																	0	0	0	0	0	0	0
Monodactylus argenteus																													1				1	0	1	0	0	0	1
Chaetodontidae (43 spp.)																																	0	0	0	0	0	0	0
Chaetodon adiergastos	1	1	1		1	1	1	1	1	1	1	1	1	1	1	1	1	1	1		1			1	1	1	1				1	1	1	1	1	1	1	1	0
Chaetodon auriga	1	1	1	1	1	1	1	1		1								1							1	1	1		1				1	1	1	1	1	1	1
Chaetodon baronessa	1	1	1	1	1		1	1	1	1	1	1	1	1	1	1	1	1	1	1	1			1	1	1	1				1	1	1	1	1	1	1	1	0
Chaetodon bennetti																													1				1	0	1	0	0	0	1
Chaetodon citrinellus	1	1	1	1	1	1	1			1	1	1	1	1	1	1	1	1	1						1		1					1	1	1	1	1	1	1	0
Chaetodon collare		1																															0	1	1	1	0	0	0
Chaetodon decussatus	1	1	1	1	1	1	1		1			1	1	1	1	1	1	1													1		1	1	1	1	1	1	0
Chaetodon ephippium	1	1	1		1													1															1	1	1	1	1	1	0
Chaetodon guentheri	1	1								1			1						1														1	1	1	1	1	1	0
Chaetodon guttatissimus	1	1		1			1																										1	1	1	1	1	0	0
Chaetodon kleinii	1	1	1	1	1	1	1	1	1	1	1	1	1	1	1	1	1	1	1	1	1			1	1	1	1				1	1	1	1	1	1	1	1	0
Chaetodon lineolatus	1	1											1																				1	1	1	1	1	0	0
Chaetodon lunula	1	1			1				1		1			1					1														1	1	1	1	1	1	0
Chaetodon lunulatus	1	1	1	1	1	1	1			1	1	1	1	1	1	1	1	1	1	1	1			1	1	1	1			1			1	1	1	1	1	1	1
Chaetodon melannotus	1	1	1	1	1	1	1	1	1	1	1	1	1	1	1	1	1	1	1					1		1			1				1	1	1	1	1	1	1
Chaetodon mertensii		1																															0	1	1	1	0	0	0
Chaetodon meyeri	1	1			1		1	1										1															1	1	1	1	1	1	0
Chaetodon ocellicaudus	1	1																						1								1	1	1	1	1	0	1	0
Chaetodon octofasciatus																														1			1	0	1	0	0	0	1
Chaetodon ornatissimus	1	1			1		1																										1	1	1	1	1	0	0
Chaetodon oxycephalus	1	1											1	1			1	1								1							1	1	1	1	1	1	0
Chaetodon punctatofasciatus	1	1								1	1																						1	1	1	1	1	0	0
Chaetodon rafflesi	1	1										1	1	1	1	1	1	1	1					1	1	1	1			1			1	1	1	1	1	1	1

table continued on next page

Appendix 3.1. *continued*

	Previous Bali	NP Surveys	Site 1	Site 2	Site 3	Site 4	Site 5	Site 7	Site 9	Site 10	Site 11	Site 12	Site 13	Site 14	Site 15	Site 16	Site 17	Site 18	Site 19	Site 20	Site 21	Site 22	Site 23	Site 24	Site 25	Site 26	Site 28	Site 33	Site 29	Site 30	Site 31	Site 32	Present Survey	Previous Surveys	Grand Total	Nusa Penida	East Bali	North Bali	Gilimanuk
Chaetodon reticulatus										1																							1	0	1	0	1	0	0
Chaetodon selene	1	1					1			1				1																	1		1	1	1	1	1	0	0
Chaetodon speculum	1	1	1	1				1		1	1				1	1	1	1	1						1		1		1				1	1	1	1	1	1	1
Chaetodon trifascialis	1	1		1		1	1	1	1	1	1	1	1	1				1	1							1	1						1	1	1	1	1	1	0
Chaetodon trifasciatus		1																1															1	1	1	1	0	1	0
Chaetodon unimaculatus	1	1				1	1	1			1	1		1																		1	1	1	1	1	1	0	0
Chaetodon vagabundus	1	1		1	1	1	1	1	1	1	1	1	1	1	1	1	1	1	1	1	1			1	1	1	1		1		1	1	1	1	1	1	1	1	1
Chaetodon xanthurus		1								1																							1	1	1	1	1	0	0
Coradion altivelis	1	1																															0	1	1	1	0	0	0
Coradion chrysozonus							1								1									1		1					1	1	1	0	1	0	1	1	0
Coradion melanopus																		1															1	0	1	0	0	1	0
Forcipiger flavissimus	1	1						1		1	1		1		1	1	1	1			1					1	1						1	1	1	1	1	1	0
Forcipiger longirostris	1	1						1		1								1								1							1	1	1	1	1	1	0
Hemitaurichthys polylepis	1	1						1		1	1			1	1		1	1						1		1	1						1	1	1	1	1	1	0
Heniochus acuminatus										1										1													1	0	1	0	1	1	0
Heniochus chrysostomus	1	1								1	1				1	1	1	1								1	1					1	1	1	1	1	1	1	0
Heniochus diphreutes	1	1						1	1				1	1					1			1	1			1			1				1	1	1	1	1	1	1
Heniochus monoceros					1		1																										1	0	1	0	1	0	0
Heniochus singularius	1	1			1					1		1		1	1						1			1		1		1	1		1		1	1	1	1	1	1	1
Heniochus varius	1	1	1	1	1			1	1	1	1		1	1	1	1	1	1						1	1	1	1		1	1	1	1	1	1	1	1	1	1	1
Pomacanthidae (21 spp.)																																	0	0	0	0	0	0	0
Apolemichthys trimaculatus	1	1						1		1		1	1	1	1			1	1							1							1	1	1	1	1	1	0
Centropyge bicolor	1	1		1			1			1	1				1	1	1	1		1	1			1		1	1		1		1	1	1	1	1	1	1	1	1
Centropyge bispinosa	1	1						1																									1	1	1	1	1	0	0
Centropyge eibli	1	1	1	1	1	1	1								1			1	1						1	1	1						1	1	1	1	1	1	0
Centropyge flavicauda	1	1						1		1			1																				1	1	1	1	1	0	0
Centropyge nox																		1								1							1	0	1	0	0	1	0
Centropyge tibicen	1	1		1	1		1	1	1	1	1	1	1	1	1	1	1	1	1	1	1			1	1	1	1		1		1	1	1	1	1	1	1	1	1

table continued on next page

Appendix 3.1. *continued*

	Previous Bali	NP Surveys	Site 1	Site 2	Site 3	Site 4	Site 5	Site 7	Site 9	Site 10	Site 11	Site 12	Site 13	Site 14	Site 15	Site 16	Site 17	Site 18	Site 19	Site 20	Site 21	Site 22	Site 23	Site 24	Site 25	Site 26	Site 28	Site 33	Site 29	Site 30	Site 31	Site 32	Present Survey	Previous Surveys	Grand Total	Nusa Penida	East Bali	North Bali	Gilimanuk
Centropyge vroliki	1	1	1	1	1	1	1	1	1			1	1	1	1	1	1	1	1	1	1			1	1	1	1				1	1	1	1	1	1	1	1	0
Chaetodontoplus melanosoma								1		1	1	1	1																				1	0	1	0	1	0	0
Chaetodontoplus mesoleucus																														1			1	0	1	0	0	0	1
Genicanthus caudivittatus		1																															0	1	1	1	0	0	0
Genicanthus lamarck	1	1						1		1								1	1						1	1	1						1	1	1	1	1	1	0
Genicanthus melanospilos	1	1								1																	1						1	1	1	1	1	1	0
Paracentropyge multifasciata													1																				1	0	1	0	1	0	0
Pomacanthus annularis			1																														1	0	1	0	1	0	0
Pomacanthus imperator	1	1						1		1		1	1	1	1	1	1	1	1	1						1	1					1	1	1	1	1	1	1	0
Pomacanthus navarchus																										1	1						1	0	1	0	0	1	0
Pomacanthus semicirculatus		1		1	1		1	1	1	1	1	1	1	1	1	1	1										1						1	1	1	1	1	1	0
Pomacanthus sexstriatus													1	1	1	1	1												1				1	0	1	0	1	1	1
Pomacanthus xanthometopon					1									1				1															1	0	1	0	1	1	0
Pygoplites diacanthus	1	1											1	1	1	1	1	1	1						1	1	1						1	1	1	1	1	1	0
Pomacentridae (96 spp.)																																	0	0	0	0	0	0	0
Abudefduf lorentzi																											1						1	0	1	0	0	1	0
Abudefduf notatus	1	1																															0	1	1	1	0	0	0
Abudefduf septemfasciatus		1										1						1						1			1						1	1	1	1	1	1	0
Abudefduf sexfasciatus	1	1	1	1	1	1	1	1	1	1	1															1	1		1				1	1	1	1	1	1	1
Abudefduf sordidus		1																1															1	1	1	1	0	1	0
Abudefduf vaigiensis	1	1	1	1	1	1	1	1	1	1	1	1		1	1	1	1	1	1	1	1			1	1	1	1				1	1	1	1	1	1	1	1	0
Amblyglyphidodon aureus	1	1								1			1	1	1	1	1	1		1	1			1	1	1	1						1	1	1	1	1	1	0
Amblyglyphidodon batunai		1																															0	1	1	1	0	0	0
Amblyglyphidodon curacao	1	1		1	1		1		1			1	1		1			1						1	1	1	1			1	1		1	1	1	1	1	1	1
Amblyglyphidodon leucogaster		1						1		1	1		1		1	1	1	1		1	1			1	1	1	1		1	1	1		1	1	1	1	1	1	1
Amblyglyphidodon ternatensis	1	1										1																	1	1			1	1	1	1	1	0	1
Amblypomacentrus breviceps	1								1													1							1	1			1	1	1	0	1	1	1
Amblypomacentrus clarus																								1									1	0	1	0	0	1	0

table continued on next page

Appendix 3.1. *continued*

	Previous Bali	NP Surveys	Site 1	Site 2	Site 3	Site 4	Site 5	Site 7	Site 9	Site 10	Site 11	Site 12	Site 13	Site 14	Site 15	Site 16	Site 17	Site 18	Site 19	Site 20	Site 21	Site 22	Site 23	Site 24	Site 25	Site 26	Site 28	Site 33	Site 29	Site 30	Site 31	Site 32	Present Survey	Previous Surveys	Grand Total	Nusa Penida	East Bali	North Bali	Gilimanuk
Amphiprion akallopisos	1	1		1											1			1															1	1	1	1	1	1	0
Amphiprion clarkii	1	1	1	1		1	1	1	1	1	1	1	1	1	1	1	1	1	1	1						1	1				1	1	1	1	1	1	1	1	0
Amphiprion frenatus																													1				1	0	1	0	0	0	1
Amphiprion melanopus		1																															0	1	1	1	0	0	0
Amphiprion ocellaris	1	1		1	1			1		1				1					1												1	1	1	1	1	1	1	1	0
Amphiprion perideraion	1	1		1				1					1	1				1							1						1		1	1	1	1	1	1	0
Amphiprion polymnus	1							1				1										1	1									1	1	1	1	0	1	1	0
Amphiprion sebae																																1	1	0	1	0	0	0	0
Chromis albicauda	1	1						1					1																				1	1	1	1	1	0	0
Chromis alpha																									1	1	1						1	0	1	0	0	1	0
Chromis amboinensis	1	1						1										1						1	1	1	1						1	1	1	1	1	1	0
Chromis analis	1	1								1			1	1	1	1	1	1								1	1						1	1	1	1	1	1	0
Chromis atripectoralis	1	1																														1	1	1	1	1	0	0	0
Chromis atripes	1	1								1	1		1		1	1	1	1		1	1			1	1	1	1						1	1	1	1	1	1	0
Chromis caudalis	1	1								1			1		1	1	1	1		1	1					1	1						1	1	1	1	1	1	0
Chromis delta	1	1											1	1	1	1	1	1		1	1			1	1	1	1						1	1	1	1	1	1	0
Chromis dimidiata		1																															0	1	1	1	0	0	0
Chromis earina																											1						1	0	1	0	0	1	0
Chromis elerae		1																1	1	1				1			1						1	1	1	1	0	1	0
Chromis lepidolepis	1	1					1	1	1	1					1	1	1	1	1	1	1			1	1	1	1				1	1	1	1	1	1	1	1	0
Chromis margaritifer	1	1	1	1	1	1	1	1	1	1	1	1	1	1	1	1	1	1	1	1	1			1	1	1	1				1	1	1	1	1	1	1	1	0
Chromis opercularis	1	1									1	1																					1	1	1	1	1	0	0
Chromis pura		1																															0	1	1	1	0	0	0
Chromis retrofasciata	1	1						1	1	1	1				1	1	1	1		1	1			1	1	1	1			1			1	1	1	1	1	1	1
Chromis scotochilopterus	1	1								1	1	1	1	1	1	1	1							1		1	1				1	1	1	1	1	1	1	1	0
Chromis sp. (70m Buyuk)		1																															0	1	1	1	0	0	0
Chromis ternatensis	1	1						1	1	1	1				1	1	1	1	1	1	1			1	1	1	1			1			1	1	1	1	1	1	1
Chromis viridis	1	1						1	1				1	1	1	1	1				1			1	1	1	1		1	1	1	1	1	1	1	1	1	1	1

table continued on next page

Appendix 3.1. *continued*

	Previous Bali	NP Surveys	Site 1	Site 2	Site 3	Site 4	Site 5	Site 7	Site 9	Site 10	Site 11	Site 12	Site 13	Site 14	Site 15	Site 16	Site 17	Site 18	Site 19	Site 20	Site 21	Site 22	Site 23	Site 24	Site 25	Site 26	Site 28	Site 33	Site 29	Site 30	Site 31	Site 32	Present Survey	Previous Surveys	Grand Total	Nusa Penida	East Bali	North Bali	Gilimanuk
Chromis weberi	1	1	1	1	1	1	1			1	1	1	1	1	1	1	1	1	1	1	1			1	1	1	1				1	1	1	1	1	1	1	1	0
Chromis xanthochira	1	1																								1							1	1	1	1	0	1	0
Chromis xanthura				1							1		1		1	1	1	1			1					1	1						1	0	1	0	1	1	0
Chrysiptera bleekeri	1	1																															0	1	1	1	0	0	0
Chrysiptera brownriggii	1	1		1						1	1			1	1	1	1	1	1								1						1	1	1	1	1	1	0
Chrysiptera glauca		1																															0	1	1	1	0	0	0
Chrysiptera rollandi	1	1														1		1		1	1			1	1	1	1			1	1	1	1	1	1	1	0	1	1
Chrysiptera springeri																															1	1	1	0	1	0	0	0	0
Chrysiptera talboti	1	1								1	1	1	1	1	1	1	1	1	1					1	1	1	1						1	1	1	1	1	1	0
Chrysiptera unimaculata	1	1													1	1	1	1	1								1						1	1	1	1	0	1	0
Dascyllus aruanus	1	1			1																			1					1	1			1	1	1	1	1	1	1
Dascyllus melanurus	1	1																											1	1			1	1	1	1	0	0	1
Dascyllus reticulatus	1	1	1	1	1		1	1	1	1	1	1	1	1	1	1	1	1	1	1	1			1	1	1	1		1	1	1	1	1	1	1	1	1	1	1
Dascyllus trimaculatus	1	1	1	1	1	1	1	1	1	1	1	1	1	1	1	1	1	1	1	1	1	1	1	1	1	1	1		1		1	1	1	1	1	1	1	1	1
Dischistodus chrysopoecilus		1																												1			1	1	1	1	0	0	1
Dischistodus melanotus																		1							1	1	1			1			1	0	1	0	0	1	1
Dischistodus perspicillatus		1																											1	1			1	1	1	1	0	0	1
Dischistodus prosopotaenia					1																								1	1			1	0	1	0	1	0	1
Hemiglyphidodon plagiometopon																														1			1	0	1	0	0	0	1
Neoglyphidodon bonang		1																															0	1	1	1	0	0	0
Neoglyphidodon crossi	1	1	1	1	1	1	1		1		1	1	1	1	1	1	1	1			1												1	1	1	1	1	1	0
Neoglyphidodon melas	1	1	1	1	1	1	1		1				1	1												1							1	1	1	1	1	1	0
Neoglyphidodon nigroris		1																			1												1	1	1	1	0	1	0
Neoglyphidodon oxyodon		1																															0	1	1	1	0	0	0
Neopomacentrus azysron																	1															1	1	0	1	0	0	1	0
Neopomacentrus cyanomos	1																			1									1			1	1	1	1	0	0	1	1
Neopomacentrus violascens	1																					1	1						1				1	1	1	0	0	1	1
Plectroglyphidodon dickii	1	1	1	1	1		1							1				1						1		1							1	1	1	1	1	1	0

table continued on next page

Appendix 3.1. *continued*

	Previous Bali	NP Surveys	Site 1	Site 2	Site 3	Site 4	Site 5	Site 7	Site 9	Site 10	Site 11	Site 12	Site 13	Site 14	Site 15	Site 16	Site 17	Site 18	Site 19	Site 20	Site 21	Site 22	Site 23	Site 24	Site 25	Site 26	Site 28	Site 33	Site 29	Site 30	Site 31	Site 32	Present Survey	Previous Surveys	Grand Total	Nusa Penida	East Bali	North Bali	Gilimanuk
Plectroglyphidodon johnstonianus	1	1									1																						1	1	1	1	1	0	0
Plectroglyphidodon lacrymatus	1	1	1	1	1		1	1	1			1		1	1	1	1	1	1					1	1	1					1	1	1	1	1	1	1	1	0
Plectroglyphidodon leucozona	1	1	1							1	1				1	1	1	1	1														1	1	1	1	1	1	0
Pomacentrus adelus		1		1	1				1			1	1		1	1	1	1						1	1	1	1				1	1	1	1	1	1	1	1	0
Pomacentrus alexanderae	1	1			1				1				1		1			1						1	1	1	1			1			1	1	1	1	1	1	1
Pomacentrus alleni		1																															0	1	1	1	0	0	0
Pomacentrus amboinensis	1	1	1	1	1			1	1	1	1	1	1	1	1	1	1	1	1	1	1			1	1	1	1		1	1	1	1	1	1	1	1	1	1	1
Pomacentrus auriventris	1	1					1	1		1	1	1	1	1	1	1	1	1	1		1			1	1	1	1		1		1	1	1	1	1	1	1	1	1
Pomacentrus bankanensis	1	1	1	1		1	1	1	1	1	1	1	1	1	1	1	1	1	1					1	1	1	1				1	1	1	1	1	1	1	1	0
Pomacentrus brachialis	1	1		1	1		1	1		1	1		1		1	1	1	1	1	1	1			1	1	1	1				1	1	1	1	1	1	1	1	0
Pomacentrus chrysurus		1	1	1		1									1	1	1	1	1					1		1	1						1	1	1	1	1	1	0
Pomacentrus coelestis	1	1	1	1	1	1	1	1	1	1	1	1	1	1	1	1	1	1	1	1	1			1	1	1	1				1	1	1	1	1	1	1	1	0
Pomacentrus grammorhynchus																														1			1	0	1	0	0	0	1
Pomacentrus lepidogenys	1	1	1	1	1	1	1	1	1		1	1	1	1	1	1	1	1	1	1	1			1	1	1	1				1	1	1	1	1	1	1	1	0
Pomacentrus melanochir	1																						1										1	1	1	0	0	1	0
Pomacentrus moluccensis	1	1	1	1	1		1	1	1	1	1	1	1	1	1	1	1	1	1	1	1			1	1	1	1			1	1	1	1	1	1	1	1	1	1
Pomacentrus nagasakiensis	1	1			1		1		1					1						1	1										1	1	1	1	1	1	1	1	0
Pomacentrus nigromarginatus	1	1								1						1	1	1		1	1			1	1	1	1						1	1	1	1	1	1	0
Pomacentrus pavo																													1				1	0	1	0	0	0	1
Pomacentrus philippinus		1	1	1	1	1	1				1			1			1									1						1	1	1	1	1	1	1	0
Pomacentrus reidi	1	1						1		1			1		1	1	1	1	1	1	1			1	1	1	1				1	1	1	1	1	1	1	1	0
Pomacentrus simsiang																													1	1			1	0	1	0	0	0	1
Pomacentrus tripunctatus																											1						1	0	1	0	0	1	0
Pomacentrus vaiuli	1	1			1	1	1	1		1			1		1	1	1	1			1			1		1	1						1	1	1	1	1	1	0
Pristotis obtrusirostris																						1											1	0	1	0	0	1	0
Stegastes fasciolatus	1	1	1															1															1	1	1	1	1	1	0
Stegastes punctatus																													1	1			1	0	1	0	0	0	1
Labridae (114 spp.)																																	0	0	0	0	0	0	0

table continued on next page

Appendix 3.1. *continued*

	Previous Bali	NP Surveys	Site 1	Site 2	Site 3	Site 4	Site 5	Site 7	Site 9	Site 10	Site 11	Site 12	Site 13	Site 14	Site 15	Site 16	Site 17	Site 18	Site 19	Site 20	Site 21	Site 22	Site 23	Site 24	Site 25	Site 26	Site 28	Site 33	Site 29	Site 30	Site 31	Site 32	Present Survey	Previous Surveys	Grand Total	Nusa Penida	East Bali	North Bali	Gilimanuk
Anampses caeruleopunctatus		1																															0	1	1	1	0	0	0
Anampses geographicus	1																																0	1	1	0	0	0	0
Anampses melanurus	1									1			1				1				1			1		1	1						1	1	1	0	1	1	0
Anampses meleagrides		1				1	1										1	1	1							1							1	1	1	1	1	1	0
Anampses twistii	1	1											1													1	1						1	1	1	1	1	1	0
Bodianus axillaris	1	1		1						1														1							1		1	1	1	1	1	1	0
Bodianus bilunulatus		1								1	1	1	1	1																			1	1	1	1	1	0	0
Bodianus bimaculatus		1												1																			1	1	1	1	1	0	0
Bodianus diana	1	1						1		1		1	1	1	1	1	1	1		1	1			1		1	1						1	1	1	1	1	1	0
Bodians izuensis		1																	1							1							1	1	1	1	0	1	0
Bodianus leucostictus		1																															0	1	1	1	0	0	0
Bodianus mesothorax	1	1	1	1	1			1	1	1	1	1	1	1	1	1	1	1			1			1	1	1	1				1	1	1	1	1	1	1	1	0
Cheilinus chlorourus	1	1	1	1	1	1	1	1	1	1	1																						1	1	1	1	1	0	0
Cheilinus fasciatus	1	1					1									1		1						1	1	1	1			1			1	1	1	1	1	1	1
Cheilinus oxycephalus	1	1						1					1				1							1		1	1						1	1	1	1	1	1	0
Cheilinus trilobatus	1	1	1				1	1	1			1			1	1	1	1	1					1	1	1	1					1	1	1	1	1	1	1	0
Cheilinus undulatus	1	1											1					1															1	1	1	1	1	1	0
Cheilio inermis	1	1			1				1				1													1							1	1	1	1	1	1	0
Choerodon anchorago	1	1					1		1																1					1			1	1	1	1	1	1	1
Choerodon zamboangae																										1							1	0	1	0	0	1	0
Cirrhilabrus brunneus								1																									1	0	1	0	1	0	0
Cirrhilabrus cf cyanopleura	1	1	1	1	1		1	1	1	1	1		1	1	1	1	1	1		1	1			1	1	1	1			1	1	1	1	1	1	1	1	1	1
Cirrhilabrus exquisitus	1	1								1																1							1	1	1	1	1	1	0
Cirrhilabrus filamentosus	1	1						1		1																							1	1	1	1	1	0	0
Cirrhilabrus flavidorsalis	1	1																								1							1	1	1	1	0	1	0
Cirrhilabrus lubbocki	1	1																1								1				1			1	1	1	1	0	1	1
Cirrhilabrus pylei		1																									1						1	1	1	1	0	1	0
Cirrhilabrus rubrimarginatus	1	1						1																		1							1	1	1	1	1	1	0

table continued on next page

Appendix 3.1. *continued*

	Previous Bali	NP Surveys	Site 1	Site 2	Site 3	Site 4	Site 5	Site 7	Site 9	Site 10	Site 11	Site 12	Site 13	Site 14	Site 15	Site 16	Site 17	Site 18	Site 19	Site 20	Site 21	Site 22	Site 23	Site 24	Site 25	Site 26	Site 28	Site 33	Site 29	Site 30	Site 31	Site 32	Present Survey	Previous Surveys	Grand Total	Nusa Penida	East Bali	North Bali	Gilimanuk
Cirrhilabrus solorensis		1																															0	1	1	1	0	0	0
Cirrhilabrus temminckii	1	1																															0	1	1	1	0	0	0
Coris aygula		1																															0	1	1	1	0	0	0
Coris batuensis	1	1						1					1											1	1				1				1	1	1	1	1	1	1
Coris dorsomacula	1	1						1	1		1	1		1																	1		1	1	1	1	1	0	0
Coris gaimardi	1	1								1		1	1		1	1	1								1	1	1				1	1	1	1	1	1	1	1	0
Coris pictoides	1																												1		1	1	1	1	1	0	0	0	1
Diproctacanthus xanthurus	1	1							1				1																	1			1	1	1	1	1	0	1
Epibulus brevis		1																															0	1	1	1	0	0	0
Epibulus insidiator	1	1		1	1		1		1	1	1		1		1	1	1							1	1	1	1		1	1			1	1	1	1	1	1	1
Gomphosus caeruleus						1	1																										1	0	1	0	1	0	0
Gomphosus varius	1	1		1	1	1	1	1	1				1	1	1	1	1							1		1	1						1	1	1	1	1	1	0
Halichoeres argus																													1				1	0	1	0	0	0	1
Halichoeres biocellatus	1	1																															0	1	1	1	0	0	0
Halichoeres chloropterus																													1	1			1	0	1	0	0	0	1
Halichoeres chrysotaenia				1	1																												1	0	1	0	1	0	0
Halichoeres chrysus	1	1					1	1				1		1	1	1	1	1	1		1			1	1	1	1						1	1	1	1	1	1	0
Halichoeres hartzfeldii	1	1												1																		1	1	1	1	1	1	0	0
Halichoeres hortulanus	1	1	1	1	1	1	1	1	1	1	1	1	1	1	1	1	1	1	1					1	1	1	1				1	1	1	1	1	1	1	1	0
Halichoeres margaritaceus	1	1		1								1	1	1	1	1	1																1	1	1	1	1	1	0
Halichoeres marginatus	1	1	1	1	1		1				1	1			1	1	1	1	1	1	1						1						1	1	1	1	1	1	0
Halichoeres melanochir		1																															0	1	1	1	0	0	0
Halichoeres melanurus															1										1					1	1	1	1	0	1	0	0	1	1
Halichoeres nebulosus	1	1																															0	1	1	1	0	0	0
Halichoeres nigrescens																							1										1	0	1	0	0	1	0
Halichoeres podostigma	1	1							1									1									1						1	1	1	1	1	1	0
Halichoeres prosopeion	1	1								1			1	1	1	1	1	1						1	1	1	1				1	1	1	1	1	1	1	1	0
Halichoeres richmondi	1	1																															0	1	1	1	0	0	0

table continued on next page

Appendix 3.1. *continued*

	Previous Bali	NP Surveys	Site 1	Site 2	Site 3	Site 4	Site 5	Site 7	Site 9	Site 10	Site 11	Site 12	Site 13	Site 14	Site 15	Site 16	Site 17	Site 18	Site 19	Site 20	Site 21	Site 22	Site 23	Site 24	Site 25	Site 26	Site 28	Site 33	Site 29	Site 30	Site 31	Site 32	Present Survey	Previous Surveys	Grand Total	Nusa Penida	East Bali	North Bali	Gilimanuk
Halichoeres scapularis	1	1			1			1	1							1		1	1					1			1		1	1			1	1	1	1	1	1	1
Halichoeres solorensis	1	1	1			1		1					1																		1	1	1	1	1	1	1	0	0
Halichoeres timorensis				1	1		1																										1	0	1	0	1	0	0
Halichoeres trimaculatus	1	1																															0	1	1	1	0	0	0
Hemigymnus fasciatus	1	1	1	1	1	1	1	1		1	1	1	1	1	1	1	1	1	1					1	1	1	1					1	1	1	1	1	1	1	0
Hemigymnus melapterus	1	1		1	1				1						1			1							1		1		1	1			1	1	1	1	1	1	1
Hologymnosus annulatus	1	1											1												1		1						1	1	1	1	1	1	0
Hologymnosus doliatus	1	1	1					1					1					1							1								1	1	1	1	1	1	0
Iniistius aneitensis	1	1																				1											1	1	1	1	0	1	0
Iniistius javanicus	1	1																															0	1	1	1	0	0	0
Iniistius melanopus												1								1		1											1	0	1	0	1	1	0
Iniistius pavo	1	1																															0	1	1	1	0	0	0
Iniistius pentadactylus	1	1														1						1											1	1	1	1	0	1	0
Iniistius tetrazona	1	1																															0	1	1	1	0	0	0
Labrichthys unilineatus	1	1		1	1		1	1	1	1	1		1	1	1	1	1	1						1	1	1	1						1	1	1	1	1	1	0
Labroides bicolor	1	1	1	1	1			1	1	1	1		1	1	1	1	1	1			1										1		1	1	1	1	1	1	0
Labroides dimidiatus	1	1	1	1	1	1	1	1	1	1	1	1	1	1	1	1	1	1		1	1			1	1	1	1			1	1	1	1	1	1	1	1	1	1
Labroides pectoralis	1	1																															0	1	1	1	0	0	0
Labropsis alleni				1											1		1																1	0	1	0	1	1	0
Labropsis manabei		1																															0	1	1	1	0	0	0
Leptojulis chrysotaenia												1	1		1											1						1	1	0	1	0	1	1	0
Leptojulis cyanopleura		1		1															1	1													1	1	1	1	1	1	0
Leptojulis polylepis ?																				1													1	0	1	0	0	1	0
Macropharyngodon negrosensis														1	1		1																1	0	1	0	1	1	0
Macropharyngodon ornatus	1	1				1	1			1	1		1													1	1					1	1	1	1	1	1	1	0
Novaculichthys taeniourus	1	1			1					1						1																	1	1	1	1	1	1	0
Oxycheilinus bimaculatus	1	1			1		1	1	1											1					1				1	1	1	1	1	1	1	1	1	1	1
Oxycheilinus digramma	1	1	1	1	1	1	1	1	1	1	1		1	1	1	1	1	1	1					1	1	1	1			1	1	1	1	1	1	1	1	1	1

table continued on next page

Appendix 3.1. *continued*

	Previous Bali	NP Surveys	Site 1	Site 2	Site 3	Site 4	Site 5	Site 7	Site 9	Site 10	Site 11	Site 12	Site 13	Site 14	Site 15	Site 16	Site 17	Site 18	Site 19	Site 20	Site 21	Site 22	Site 23	Site 24	Site 25	Site 26	Site 28	Site 33	Site 29	Site 30	Site 31	Site 32	Present Survey	Previous Surveys	Grand Total	Nusa Penida	East Bali	North Bali	Gilimanuk
Oxycheilinus unifasciatus													1																				1	0	1	0	1	0	0
Paracheilinus sp.		1																															0	1	1	1	0	0	0
Paracheilinus filamentosus		1																		1													1	1	1	1	0	1	0
Paracheilinus flavianalis	1	1						1					1		1			1							1	1			1		1	1	1	1	1	1	1	1	1
Pseudocheilinus evanidus	1	1						1					1																				1	1	1	1	1	0	0
Pseudocheilinus hexataenia	1	1	1	1	1		1	1	1	1	1		1					1	1	1	1			1	1	1	1						1	1	1	1	1	1	0
Pseudocheilinus octotaenia	1	1																															0	1	1	1	0	0	0
Pseudocoris bleekeri														1																			1	0	1	0	1	0	0
Pseudocoris heteroptera	1	1																								1							1	1	1	1	0	1	0
Pseudocoris yamashiroi	1	1																															0	1	1	1	0	0	0
Pseudodax moluccanus	1	1								1	1		1	1												1	1				1	1	1	1	1	1	1	1	0
Pseudojuloides cerasinus								1																									1	0	1	0	1	0	0
Pseudojuloides kaleidos	1	1																															0	1	1	1	0	0	0
Pseudojuloides mesostigma																											1						1	0	1	0	0	1	0
Pseudojuloides severnsi								1																									1	0	1	0	1	0	0
Pteragogus cryptus													1																				1	0	1	0	1	0	0
Pteragogus enneacanthus	1	1																															0	1	1	1	0	0	0
Stethojulis bandanensis	1	1			1										1	1	1								1								1	1	1	1	1	1	0
Stethojulis interrupta	1	1					1	1	1																1								1	1	1	1	1	1	0
Stethojulis strigiventer		1																											1	1			1	1	1	1	0	0	1
Stethojulis trilineata	1	1					1		1						1	1	1																1	1	1	1	1	1	0
Terelabrus rubrovittatus		1																															0	1	1	1	0	0	0
Thalassoma amblycephalus	1	1	1	1	1	1	1	1	1	1	1	1	1	1	1	1	1	1	1					1	1	1	1				1	1	1	1	1	1	1	1	0
Thalassoma hardwicke	1	1	1	1	1		1	1	1	1	1	1	1	1	1	1	1	1	1					1	1	1	1			1	1	1	1	1	1	1	1	1	1
Thalassoma jansenii	1	1	1	1	1	1	1		1	1	1	1	1	1	1	1	1	1	1								1				1	1	1	1	1	1	1	1	0
Thalassoma lunare	1	1	1	1	1	1	1	1	1	1	1	1	1	1				1	1	1	1			1	1	1	1		1	1	1	1	1	1	1	1	1	1	1
Thalassoma purpureum	1	1												1																			1	1	1	1	1	0	0
Thalassoma quinquevittatum	1	1																															0	1	1	1	0	0	0

table continued on next page

Appendix 3.1. *continued*

	Previous Bali	NP Surveys	Site 1	Site 2	Site 3	Site 4	Site 5	Site 7	Site 9	Site 10	Site 11	Site 12	Site 13	Site 14	Site 15	Site 16	Site 17	Site 18	Site 19	Site 20	Site 21	Site 22	Site 23	Site 24	Site 25	Site 26	Site 28	Site 33	Site 29	Site 30	Site 31	Site 32	Present Survey	Previous Surveys	Grand Total	Nusa Penida	East Bali	North Bali	Gilimanuk
Thalassoma trilobatum	1	1												1																			1	1	1	1	1	0	0
Wetmorella nigropinnata		1																															0	1	1	1	0	0	0
Scaridae (24 spp.)																																	0	0	0	0	0	0	0
Bolbometopon muricatum										1				1				1						1									1	0	1	0	1	1	0
Calotomus carolinus	1	1			1	1		1																									1	1	1	1	1	0	0
Cetoscarus ocellatus	1	1																1						1			1						1	1	1	1	0	1	0
Chlorurus bleekeri	1	1																							1	1	1						1	1	1	1	0	1	0
Chlorurus capistratoides		1						1			1				1	1		1	1					1									1	1	1	1	1	1	0
Chlorurus microrhinos																											1						1	0	1	0	0	1	0
Chlorurus sordidus	1	1		1	1			1	1	1	1			1											1	1	1				1	1	1	1	1	1	1	1	0
Leptoscarus vaigiensis		1																															0	1	1	1	0	0	0
Scarus dimidatus		1																							1		1		1	1			1	1	1	1	0	1	1
Scarus festivus	1	1																															0	1	1	1	0	0	0
Scarus flavipectoralis	1	1								1	1																						1	1	1	1	1	0	0
Scarus forsteni	1	1															1	1															1	1	1	1	0	1	0
Scarus frenatus	1	1															1																1	1	1	1	0	1	0
Scarus ghobban	1	1		1	1	1	1	1	1					1												1			1		1	1	1	1	1	1	1	1	1
Scarus niger	1	1		1	1					1				1	1	1	1	1	1					1	1	1	1						1	1	1	1	1	1	0
Scarus oviceps		1						1	1	1	1							1						1		1	1						1	1	1	1	1	1	0
Scarus prasiognathos								1						1													1						1	0	1	0	1	1	0
Scarus psittacus	1	1	1	1	1	1																											1	1	1	1	1	0	0
Scarus quoyi		1							1							1	1	1							1	1	1					1	1	1	1	1	1	1	0
Scarus rivulatus		1														1	1										1					1	1	1	1	1	0	1	0
Scarus rubroviolaceus	1	1					1					1		1	1	1	1		1					1			1						1	1	1	1	1	1	0
Scarus schlegeli	1	1														1	1																1	1	1	1	0	1	0
Scarus spinus	1	1							1								1	1						1			1						1	1	1	1	1	1	0
Scarus tricolor		1										1		1	1	1	1							1		1	1						1	1	1	1	1	1	0
Trichonotidae 3 spp.)																																	0	0	0	0	0	0	0

table continued on next page

Appendix 3.1. *continued*

	Previous Bali	NP Surveys	Site 1	Site 2	Site 3	Site 4	Site 5	Site 7	Site 9	Site 10	Site 11	Site 12	Site 13	Site 14	Site 15	Site 16	Site 17	Site 18	Site 19	Site 20	Site 21	Site 22	Site 23	Site 24	Site 25	Site 26	Site 28	Site 33	Site 29	Site 30	Site 31	Site 32	Present Survey	Previous Surveys	Grand Total	Nusa Penida	East Bali	North Bali	Gilimanuk
Pteropsaron springeri																		1			1			1		1							1	0	1	0	0	1	0
Trichonotus elegans		1																															0	1	1	1	0	0	0
Trichonotus setiger	1											1				1	1																1	1	1	0	1	1	0
Creediidae (1 spp.)																																	0	0	0	0	0	0	0
Limnichthys nitidus		1										1					1																1	1	1	1	1	1	0
Pinguipedidae (10 spp.)																																	0	0	0	0	0	0	0
Parapercis bimacula	1																																0	1	1	0	0	0	0
Parapercis clathrata	1	1												1	1	1	1	1							1						1	1	1	1	1	1	1	1	0
Parapercis cylindrica	1	1	1	1	1				1							1	1								1				1	1	1	1	1	1	1	1	1	1	1
Parapercis flavolineata																									1	1							1	0	1	0	0	1	0
Parapercis hexophtalma	1	1		1	1																			1	1	1	1						1	1	1	1	1	1	0
Parapercis maculata	1																																0	1	1	0	0	0	0
Parapercis millepunctata	1	1				1									1	1	1		1	1	1					1							1	1	1	1	1	1	0
Parapercis schauinslandii	1	1						1	1										1														1	1	1	1	1	1	0
Parapercis sp. (photos)							1	1	1							1															1		1	0	1	0	1	1	0
Parapercis tetracantha	1	1											1	1	1	1	1	1	1					1									1	1	1	1	1	1	0
Trypterygiidae 14 spp,)																																	0	0	0	0	0	0	0
Ceratobregma helenae		1																															0	1	1	1	0	0	0
Enneapterygius flavoccipitis																					1												1	0	1	0	0	1	0
Enneapterygius hemimelas	1	1																															0	1	1	1	0	0	0
Enneapterygius similis										1																							1	0	1	0	1	0	0
Enneapterygius tutuilae															1				1								1						1	0	1	0	0	1	0
Enneapterygius sp 1 (photo)														1																			1	0	1	0	1	0	0
Helcogramma kranos?		1																															0	1	1	1	0	0	0
Helcogramma randalli														1																			1	0	1	0	1	0	0
Helcogramma rhinoceros												1																					1	0	1	0	1	0	0
Helcogramma sp. 1 (dark saddles)		1																															0	1	1	1	0	0	0
Helcogramma sp. 2 (photo)															1				1														1	0	1	0	0	1	0

table continued on next page

Appendix 3.1. *continued*

	Previous Bali	NP Surveys	Site 1	Site 2	Site 3	Site 4	Site 5	Site 7	Site 9	Site 10	Site 11	Site 12	Site 13	Site 14	Site 15	Site 16	Site 17	Site 18	Site 19	Site 20	Site 21	Site 22	Site 23	Site 24	Site 25	Site 26	Site 28	Site 33	Site 29	Site 30	Site 31	Site 32	Present Survey	Previous Surveys	Grand Total	Nusa Penida	East Bali	North Bali	Gilimanuk
Helcogramma striatum	1	1		1		1		1		1	1	1	1	1	1	1	1	1	1														1	1	1	1	1	1	0
Norfolkia brachylepis		1																															0	1	1	1	0	0	0
Ucla xenogrammus	1	1																															0	1	1	1	0	0	0
Clinidae (1 spp.)																																	0	0	0	0	0	0	0
Springeratus xanthosoma		1																															0	1	1	1	0	0	0
Blenniidae (27 spp.)																																	0	0	0	0	0	0	0
Aspidontus taeniatus																							1										1	0	1	0	0	1	0
Atrosalarias fuscus		1																								1				1			1	1	1	1	0	1	1
Blenniella chrysospilos																											1						1	0	1	0	0	1	0
Cirripectes auritus		1																															0	1	1	1	0	0	0
Cirripectes filamentosus			1																														1	0	1	0	1	0	0
Cirripectes polyzona											1			1																			1	0	1	0	1	0	0
Ecsenius bathi		1																															0	1	1	1	0	0	0
Ecsenius bicolor		1						1			1			1			1	1	1						1		1						1	1	1	1	1	1	0
Ecsenius namiyei																								1									1	0	1	0	0	1	0
Ecsenius ops																								1									1	0	1	0	0	1	0
Ecsenius shirleyae																									1		1						1	0	1	0	0	1	0
Ecsenius yaeyamaenis																1																	1	0	1	0	0	1	0
Entomacrodus decussatus																											1						1	0	1	0	0	1	0
Entomacrodus vermiculatus														1																			1	0	1	0	1	0	0
Istiblennius edentulus																											1						1	0	1	0	0	1	0
Meiacanthus cf abditus																										1							1	0	1	0	0	1	0
Meiacanthus abruptus																														1			1	0	1	0	0	0	1
Meiacanthus atrodorsalis															1			1			1			1	1	1					1		1	0	1	0	0	1	0
Meiacanthus cyanopterus																			1														1	0	1	0	0	1	0
Meiacanthus grammistes																								1	1								1	0	1	0	0	1	0
Nannosalarias nativitatus										1	1			1																			1	0	1	0	1	0	0
Petroscirtes breviceps		1							1														1	1					1				1	1	1	1	1	1	1

table continued on next page

Appendix 3.1. *continued*

	Previous Bali	NP Surveys	Site 1	Site 2	Site 3	Site 4	Site 5	Site 7	Site 9	Site 10	Site 11	Site 12	Site 13	Site 14	Site 15	Site 16	Site 17	Site 18	Site 19	Site 20	Site 21	Site 22	Site 23	Site 24	Site 25	Site 26	Site 28	Site 33	Site 29	Site 30	Site 31	Site 32	Present Survey	Previous Surveys	Grand Total	Nusa Penida	East Bali	North Bali	Gilimanuk
Petroscirtes variabilis		1																															0	1	1	1	0	0	0
Plagiotremus rhinorhynchus	1	1	1	1	1			1		1			1	1	1	1	1	1							1	1	1		1				1	1	1	1	1	1	1
Plagiotremus tapeinosoma														1					1														1	0	1	0	1	1	0
Salarias fasciatus		1																															0	1	1	1	0	0	0
Salarias guttatus		1																									1						1	1	1	1	0	1	0
Callionymidae (7 spp.)																																	0	0	0	0	0	0	0
Callionymus filamentosus																						1											1	0	1	0	0	1	0
Callionymus superbus (photo - Tul)	1																																0	1	1	0	0	0	0
Callionymus sp. 1 (photo)																			1														1	0	1	0	0	1	0
Callionymus sp. 2 (photo)																					1												1	0	1	0	0	1	0
Dactylopus dactylopus																													1				1	0	1	0	0	0	1
Synchiropus ocellatus		1																															0	1	1	1	0	0	0
Synchiropus tudorjonesi																		1															1	0	1	0	0	1	0
Gobiidae (84 spp.)																																	0	0	0	0	0	0	0
Amblyeleotris fasciata	1	1													1	1	1		1													1	1	1	1	1	0	1	0
Amblyeleotris fontanesii																							1										1	0	1	0	0	1	0
Amblyeleotris guttata																										1						1	1	0	1	0	0	1	0
Amblyeleotris periophthalma																																1	1	0	1	0	0	0	0
Amblyeleotris steinitzi	1																																0	1	1	0	0	0	0
Amblyeleotris yanoi	1																1																1	1	1	0	0	1	0
Amblygobius nocturnus																					1									1			1	0	1	0	0	1	1
Amblygobius phalaena																											1		1				1	0	1	0	0	1	1
Asterropteryx ensifera	1								1						1	1	1													1			1	1	1	0	1	1	1
Asterropteryx striata																		1	1		1			1						1			1	0	1	0	0	1	1
Bryaninops amplus										1														1									1	0	1	0	1	1	0
Bryaninops tigris?		1																															0	1	1	1	0	0	0
Cryptocentrus caeruleomaculatus					1																												1	0	1	0	1	0	0
Cryptocentrus inexplicatus																													1				1	0	1	0	0	0	1

table continued on next page

Appendix 3.1. *continued*

	Previous Bali	NP Surveys	Site 1	Site 2	Site 3	Site 4	Site 5	Site 7	Site 9	Site 10	Site 11	Site 12	Site 13	Site 14	Site 15	Site 16	Site 17	Site 18	Site 19	Site 20	Site 21	Site 22	Site 23	Site 24	Site 25	Site 26	Site 28	Site 33	Site 29	Site 30	Site 31	Site 32	Present Survey	Previous Surveys	Grand Total	Nusa Penida	East Bali	North Bali	Gilimanuk
Cryptocentrus leptocephalus																													1				1	0	1	0	0	0	1
Cryptocentrus leucostictus				1																													1	0	1	0	1	0	0
Cryptocentrus strigilliceps			1		1																								1	1			1	0	1	0	1	0	1
Ctenogobiops pomastictus			1	1	1																				1								1	0	1	0	1	1	0
Drombus species 1			1																														1	0	1	0	1	0	0
Drombus species 2																							1										1	0	1	0	0	1	0
Eviota guttata	1	1					1				1	1		1					1						1	1	1						1	1	1	1	1	1	0
Eviota prasites	1																																0	1	1	0	0	0	0
Eviota punctulata														1																			1	0	1	0	1	0	0
Eviota queenslandica?		1																															0	1	1	1	0	0	0
Eviota rubrisparsa																	1							1			1						1	0	1	0	0	1	0
Eviota sebreei		1									1				1	1	1	1	1						1								1	1	1	1	1	1	0
Eviota sigillata																1								1									1	0	1	0	0	1	0
Eviota sp. 1 (red head)																									1								1	0	1	0	0	1	0
Exyrias akihito																									1								1	0	1	0	0	1	0
Exyrias ferrarisi																													1	1			1	0	1	0	0	0	1
Fusigobius duospilus		1																															0	1	1	1	0	0	0
Fusigobius inframaculatus	1	1												1			1			1									1		1		1	1	1	1	1	1	1
Fusigobius melacron	1																														1		1	1	1	0	0	0	0
Fusigobius neophytus		1																	1														1	1	1	1	0	1	0
Fusigobius signipinnis																								1	1						1		1	0	1	0	0	1	0
Gladiogobius ensifer																														1			1	0	1	0	0	0	1
Gnatholepis cauerensis	1	1																	1	1				1	1					1			1	1	1	1	0	1	1
Gobiid sp. 1 (photo - Tulamben)	1																																0	1	1	0	0	0	0
Gobiid sp. 2 (photo - Tulamben)	1																																0	1	1				
Gobiodon citrinus		1																															0	1	1	1	0	0	0
Gobiodon prolixus																										1							1	0	1	0	0	1	0
Gobiodon quinquestrigatus	1																																0	1	1	0	0	0	0

table continued on next page

Appendix 3.1. *continued*

	Previous Bali	NP Surveys	Site 1	Site 2	Site 3	Site 4	Site 5	Site 7	Site 9	Site 10	Site 11	Site 12	Site 13	Site 14	Site 15	Site 16	Site 17	Site 18	Site 19	Site 20	Site 21	Site 22	Site 23	Site 24	Site 25	Site 26	Site 28	Site 33	Site 29	Site 30	Site 31	Site 32	Present Survey	Previous Surveys	Grand Total	Nusa Penida	East Bali	North Bali	Gilimanuk
Gobiodon sp.1 (dark with 2 blue bars)	1																																0	1	1	0	0	0	0
Gobiodon sp.2 (br with many blue bars)	1																																0	1	1	0	0	0	0
Grallenia baliensis																	1				1												1	0	1	0	0	1	0
Hazeus otakii																						1	1										1	0	1	0	0	1	0
Istigobius decoratus	1	1		1	1								1	1	1	1	1	1	1	1	1				1				1	1	1	1	1	1	1	1	1	1	1
Istigobius sp. 1 (70 m photo)																									1								1	0	1	0	0	1	0
Istigobius spence							1																								1		1	0	1	0	1	0	0
Mahidolia mystacinus																			1				1										1	0	1	0	0	1	0
Oplopomus caninoides																						1	1										1	0	1	0	0	1	0
Oplopomus oplopomus			1																										1	1			1	0	1	0	1	0	1
Paragobiodon xanthosoma					1																												1	0	1	0	1	0	0
Pleurosicya annadalei																										1							1	0	1	0	0	1	0
Pleurosicya labiata																															1		1	0	1	0	0	0	0
Pleurosicya mossambica																			1	1													1	0	1	0	0	1	0
Priolepis cinctus	1	1			1									1									1										1	1	1	1	1	1	0
Priolepis compita		1																															0	1	1	1	0	0	0
Priolepis nuchifasciatus		1																															0	1	1	1	0	0	0
Priolepis semidoliatus		1																															0	1	1	1	0	0	0
Priolepis sp. 1 (broad yellow bars - 70 m)		1																															0	1	1	1	0	0	0
Priolepis sp. 2 (photo)										1																1							1	0	1	0	1	1	0
Tomiyamichthys oni	1																																0	1	1	0	0	0	0
Trimma annosum							1																										1	0	1	0	1	0	0
Trimma benjamini		1																															0	1	1	1	0	0	0
Trimma fucatum							1												1														1	0	1	0	1	1	0
Trimma halonevum	1																1		1	1			1		1	1							1	1	1	0	0	1	0
Trimma kudoi																								1									1	0	1	0	0	1	0
Trimma imaii		1																															0	1	1	1	0	0	0

table continued on next page

Appendix 3.1. *continued*

	Previous Bali	NP Surveys	Site 1	Site 2	Site 3	Site 4	Site 5	Site 7	Site 9	Site 10	Site 11	Site 12	Site 13	Site 14	Site 15	Site 16	Site 17	Site 18	Site 19	Site 20	Site 21	Site 22	Site 23	Site 24	Site 25	Site 26	Site 28	Site 33	Site 29	Site 30	Site 31	Site 32	Present Survey	Previous Surveys	Grand Total	Nusa Penida	East Bali	North Bali	Gilimanuk
Trimma macrophthalma	1	1																								1							1	1	1	1	0	1	0
Trimma maiandros	1																																0	1	1	0	0	0	0
Trimma nomurai																			1					1									1	0	1	0	0	1	0
Trimma okinawae		1					1							1	1						1					1							1	1	1	1	1	1	0
Trimma stobbsi		1																															0	1	1	1	0	0	0
Trimma taylori	1															1								1			1						1	1	1	0	0	1	0
Trimma tevegae																		1						1	1	1	1						1	0	1	0	0	1	0
Trimma yanoi																					1			1									1	0	1	0	0	1	0
Tryssogobius sarah																		1															1	0	1	0	0	1	0
Valenciennea helsdingenii	1	1													1	1	1			1				1		1							1	1	1	1	0	1	0
Valenciennea puellaris	1	1													1	1	1	1			1			1									1	1	1	1	0	1	0
Valenciennea sexguttata		1	1													1		1	1							1			1				1	1	1	1	1	1	1
Valenciennea strigata												1	1	1	1	1	1	1			1			1	1	1							1	0	1	0	1	1	0
Vanderhorstia lanceolata																			1														1	0	1	0	0	1	0
Vanderhorstia species 1																							1										1	0	1	0	0	1	0
Xenisthmidae (1 spp.)																																	0	0	0	0	0	0	0
Xenisthmus polyzonatus?		1																															0	1	1	1	0	0	0
Microdesmidae (2 spp.)																																	0	0	0	0	0	0	0
Gunnellichthys curiosus								1									1																1	0	1	0	1	1	0
Gunnellichthys viridescens																					1			1									1	0	1	0	0	1	0
Ptereleotridae (8 spp.)																																	0	0	0	0	0	0	0
Nemateleotris decora		1								1															1	1	1						1	1	1	1	1	1	0
Nemateleotris magnifica		1									1		1					1								1							1	1	1	1	1	1	0
Ptereleotris brachyptera																				1				1									1	0	1	0	0	1	0
Ptereleotris evides	1	1			1	1	1	1		1	1		1	1	1	1	1	1							1	1	1						1	1	1	1	1	1	0
Ptereleotris grammica		1																															0	1	1	1	0	0	0
Ptereleotris hanae	1																					1											1	1	1	0	0	1	0

table continued on next page

Appendix 3.1. *continued*

	Previous Bali	NP Surveys	Site 1	Site 2	Site 3	Site 4	Site 5	Site 7	Site 9	Site 10	Site 11	Site 12	Site 13	Site 14	Site 15	Site 16	Site 17	Site 18	Site 19	Site 20	Site 21	Site 22	Site 23	Site 24	Site 25	Site 26	Site 28	Site 33	Site 29	Site 30	Site 31	Site 32	Present Survey	Previous Surveys	Grand Total	Nusa Penida	East Bali	North Bali	Gilimanuk
Ptereleotris heteroptera	1	1						1		1			1		1	1	1																1	1	1	1	1	1	0
Ptereleotris rubristigma												1				1								1									1	0	1	0	1	1	0
Ephippidae (4 spp.)																																	0	0	0	0	0	0	0
Platax boersi				1	1	1		1					1												1	1	1		1				1	0	1	0	1	1	1
Platax orbicularis					1								1										1		1								1	0	1	0	1	1	0
Platax pinnatus						1																			1								1	0	1	0	1	1	0
Platax teira	1																																0	1	1	0	0	0	0
Siganidae (13 spp.)																																	0	0	0	0	0	0	0
Siganus argenteus	1	1						1									1	1	1					1		1						1	1	1	1	1	1	1	0
Siganus canaliculatus																										1	1						1	0	1	0	0	1	0
Siganus corallinus	1	1										1	1		1	1	1	1	1					1									1	1	1	1	1	1	0
Siganus guttatus				1	1	1		1									1								1	1			1			1	1	0	1	0	1	1	1
Siganus labyrinthodes																		1														1	1	0	1	0	0	1	0
Siganus margaritifer																		1	1							1							1	0	1	0	0	1	0
Siganus puellus	1							1		1			1			1		1	1					1		1				1		1	1	1	1	0	1	1	1
Siganus punctatissimus																		1															1	0	1	0	0	1	0
Siganus punctatus																	1								1		1			1			1	0	1	0	0	1	1
Siganus spinus	1	1				1			1						1	1	1																1	1	1	1	1	1	0
Siganus vermiculatus																	1																1	0	1	0	0	1	0
Siganus virgatus	1				1										1	1	1		1					1	1	1	1			1			1	1	1	0	1	1	1
Siganus vulpinus	1												1					1			1			1		1	1						1	1	1	0	1	1	0
Zanclidae (1 spp.)																																	0	0	0	0	0	0	0
Zanclus cornutus	1	1	1	1	1	1	1	1	1	1	1	1	1	1	1	1	1	1	1	1	1			1	1	1	1				1	1	1	1	1	1	1	1	0
Acanthuridae (39 spp.)																																	0	0	0	0	0	0	0
Acanthurus barine	1	1	1									1									1												1	1	1	1	1	1	0
Acanthurus blochii															1	1	1	1													1	1	1	0	1	0	0	1	0
Acanthurus dussumieri	1	1				1							1				1																1	1	1	1	1	1	0
Acanthurus leucocheilus	1	1						1		1		1																			1		1	1	1	1	1	0	0

table continued on next page

Appendix 3.1. *continued*

	Previous Bali	NP Surveys	Site 1	Site 2	Site 3	Site 4	Site 5	Site 7	Site 9	Site 10	Site 11	Site 12	Site 13	Site 14	Site 15	Site 16	Site 17	Site 18	Site 19	Site 20	Site 21	Site 22	Site 23	Site 24	Site 25	Site 26	Site 28	Site 33	Site 29	Site 30	Site 31	Site 32	Present Survey	Previous Surveys	Grand Total	Nusa Penida	East Bali	North Bali	Gilimanuk
Acanthurus leucosternon	1	1						1							1																		1	1	1	1	1	1	0
Acanthurus lineatus	1	1	1	1	1	1	1	1		1	1	1	1	1	1	1	1	1	1							1	1				1	1	1	1	1	1	1	1	0
Acanthurus maculiceps		1													1										1		1						1	1	1	1	0	1	0
Acanthurus mata	1	1					1	1	1	1	1	1						1	1	1	1			1		1	1				1	1	1	1	1	1	1	1	0
Acanthurus nigricans	1	1				1	1	1		1	1	1	1	1	1	1	1	1	1														1	1	1	1	1	1	0
Acanthurus nigricauda	1	1		1	1	1	1	1	1			1	1	1	1	1	1	1	1					1		1	1		1		1	1	1	1	1	1	1	1	1
Acanthurus nigrofuscus	1	1	1	1	1	1	1	1	1	1	1	1	1	1	1	1	1	1													1	1	1	1	1	1	1	1	0
Acanthurus olivaceus	1	1		1				1	1		1	1			1	1	1	1	1													1	1	1	1	1	1	1	0
Acanthurus pyroferus	1	1		1				1		1	1	1	1	1	1	1	1	1	1	1	1			1	1	1	1				1	1	1	1	1	1	1	1	0
Acanthurus tennentii	1																																0	1	1	0	0	0	0
Acanthurus thompsoni	1	1								1						1		1								1	1						1	1	1	1	1	1	0
Acanthurus triostegus	1	1													1	1	1										1						1	1	1	1	0	1	0
Acanthurus tristis	1	1				1	1	1				1																				1	1	1	1	1	1	0	0
Acanthurus xanthopterus	1	1		1											1				1				1						1		1	1	1	1	1	1	1	1	1
Ctenochaetus binotatus	1	1																		1	1			1	1	1	1		1				1	1	1	1	0	1	1
Ctenochaetus cyanocheilus	1	1																															0	1	1	1	0	0	0
Ctenochaetus striatus	1	1	1	1	1	1	1	1	1	1	1	1	1	1	1	1	1	1	1					1	1	1	1				1	1	1	1	1	1	1	1	0
Ctenochaetus truncatus		1									1																				1	1	1	1	1	1	1	0	0
Naso annulatus	1	1																															0	1	1	1	0	0	0
Naso brachycentron	1	1		1				1			1		1	1	1	1	1	1	1						1	1							1	1	1	1	1	1	0
Naso brevirostris	1	1			1	1		1						1	1			1	1								1		1		1		1	1	1	1	1	1	1
Naso caeruleacauda																			1														1	0	1	0	0	1	0
Naso elegans	1	1																															0	1	1	1	0	0	0
Naso hexacanthus	1	1				1		1		1					1	1	1	1	1							1	1				1		1	1	1	1	1	1	0
Naso lituratus	1	1		1						1			1	1	1	1	1	1	1							1							1	1	1	1	1	1	0
Naso lopezi	1	1																						1									1	1	1	1	0	1	0
Naso minor	1	1								1	1		1		1	1	1	1			1					1							1	1	1	1	1	1	0
Naso reticulatus	1																																0	1	1	0	0	0	0

table continued on next page

Appendix 3.1. *continued*

	Previous Bali	NP Surveys	Site 1	Site 2	Site 3	Site 4	Site 5	Site 7	Site 9	Site 10	Site 11	Site 12	Site 13	Site 14	Site 15	Site 16	Site 17	Site 18	Site 19	Site 20	Site 21	Site 22	Site 23	Site 24	Site 25	Site 26	Site 28	Site 33	Site 29	Site 30	Site 31	Site 32	Present Survey	Previous Surveys	Grand Total	Nusa Penida	East Bali	North Bali	Gilimanuk
Naso thynnoides	1	1							1								1	1	1							1							1	1	1	1	1	1	0
Naso unicornis	1	1								1				1			1	1	1														1	1	1	1	1	1	0
Naso vlamingii	1	1	1		1			1		1		1	1	1	1	1	1	1	1	1	1				1	1	1		1				1	1	1	1	1	1	1
Paracanthurus hepatus		1						1										1															1	1	1	1	1	1	0
Prionurus chrysurus		1			1			1				1	1	1																			1	1	1	1	1	0	0
Zebrasoma scopas	1	1	1	1	1	1	1	1	1	1	1	1	1	1	1	1	1	1	1					1		1							1	1	1	1	1	1	0
Zebrasoma veliferum																	1	1					1										1	0	1	0	0	1	0
Sphyraenidae (4 spp.)																																	0	0	0	0	0	0	0
Sphyraena barracuda		1						1								1			1														1	1	1	1	1	1	0
Sphyraena jello													1				1																1	0	1	0	1	1	0
Sphyraena obtusata																				1		1											1	0	1	0	0	1	0
Sphyraena qenie					1																												1	0	1	0	1	0	0
Scombridae (6 spp.)																																	0	0	0	0	0	0	0
Euthynnus affinis																										1							1	0	1	0	0	1	0
Grammatorcynus bilineatus		1											1																				1	1	1	1	1	0	0
Gymnosarda unicolor		1											1					1						1									1	1	1	1	1	1	0
Rastrelliger kanagurta																1										1	1						1	0	1	0	0	1	0
Scomberomorus commrsonnianus		1													1																		1	1	1	1	0	1	0
Thunnus albacares		1																															0	1	1	1	0	0	0
Bothidae (3 spp.)																																	0	0	0	0	0	0	0
Asterorhombus intermedius																							1										1	0	1	0	0	1	0
Bothus mancus	1	1	1		1											1																	1	1	1	1	1	1	0
Bothus pantherinus		1												1		1						1	1										1	1	1	1	1	1	0
Soleidae (6 spp.)																																	0	0	0	0	0	0	0
Aseraggodes chapleaui		1																															0	1	1	1	0	0	0
Aseraggodes suzimotoi		1																															0	1	1	1	0	0	0
Brachirus marmoratus		1																															0	1	1	1	0	0	0
Liachirus melanospilos																							1										1	0	1	0	0	1	0

table continued on next page

Appendix 3.1. *continued*

	Previous Bali	NP Surveys	Site 1	Site 2	Site 3	Site 4	Site 5	Site 7	Site 9	Site 10	Site 11	Site 12	Site 13	Site 14	Site 15	Site 16	Site 17	Site 18	Site 19	Site 20	Site 21	Site 22	Site 23	Site 24	Site 25	Site 26	Site 28	Site 33	Site 29	Site 30	Site 31	Site 32	Present Survey	Previous Surveys	Grand Total	Nusa Penida	East Bali	North Bali	Gilimanuk
Pardachirus pavoninus																1																	1	0	1	0	0	1	0
Soleichthys heterorhinos			1																														1	0	1	0	1	0	0
Samaridae (1 spp.)																																	0	0	0	0	0	0	0
Samariscus triocellatus		1																															0	1	1	1	0	0	0
Balistidae (17 spp.)																																	0	0	0	0	0	0	0
Abalistes stellatus																1						1	1										1	0	1	0	0	1	0
Balistapus undulatus	1	1	1	1	1	1	1	1	1			1	1	1	1	1	1	1	1	1	1			1	1	1	1				1	1	1	1	1	1	1	1	0
Balistoides conspicillum	1	1						1					1		1	1																	1	1	1	1	1	1	0
Balistoides viridescens	1	1	1	1				1	1				1	1	1	1	1	1	1							1	1						1	1	1	1	1	1	0
Canthidermis maculatus																1															1	1	1	0	1	0	0	1	0
Melichthys indicus		1											1																				1	1	1	1	1	0	0
Melichthys niger	1	1					1			1			1				1																1	1	1	1	1	1	0
Melichthys vidua	1	1				1	1	1		1		1	1	1	1	1	1	1						1	1	1	1						1	1	1	1	1	1	0
Odonus niger	1	1						1		1	1		1	1	1	1	1	1			1			1	1	1							1	1	1	1	1	1	0
Pseudobalistes flavimarginatus	1	1		1	1			1											1										1				1	1	1	1	1	1	1
Pseudobalistes fuscus		1																															0	1	1	1	0	0	0
Rhinecanthus rectangulus	1	1																															0	1	1	1	0	0	0
Rhinecanthus verrucosus		1													1	1	1	1								1	1						1	1	1	1	0	1	0
Sufflamen bursa	1	1						1		1	1		1	1	1	1	1	1			1			1	1	1					1	1	1	1	1	1	1	1	0
Sufflamen chrysopterus	1	1	1	1	1		1	1	1	1	1		1	1	1	1	1	1						1	1						1	1	1	1	1	1	1	1	0
Sufflamen frenatus		1																															0	1	1	1	0	0	0
Xanthichthys auromarginatus		1								1			1																				1	1	1	1	1	0	0
Monacanthidae (14 spp.)																																	0	0	0	0	0	0	0
Acreichthys tomentosus																						1	1						1				1	0	1	0	0	1	1
Aluterus scriptus	1	1						1	1	1	1		1	1							1						1					1	1	1	1	1	1	1	0
Amanses scopas	1	1											1	1	1	1	1	1			1			1		1	1						1	1	1	1	1	1	0
Cantherhines dumerilii	1	1		1				1		1	1			1	1	1	1	1															1	1	1	1	1	1	0
Cantherhines fronticinctua	1	1		1						1								1															1	1	1	1	1	1	0

table continued on next page

Appendix 3.1. *continued*

	Previous Bali	NP Surveys	Site 1	Site 2	Site 3	Site 4	Site 5	Site 7	Site 9	Site 10	Site 11	Site 12	Site 13	Site 14	Site 15	Site 16	Site 17	Site 18	Site 19	Site 20	Site 21	Site 22	Site 23	Site 24	Site 25	Site 26	Site 28	Site 33	Site 29	Site 30	Site 31	Site 32	Present Survey	Previous Surveys	Grand Total	Nusa Penida	East Bali	North Bali	Gilimanuk
Cantherhines pardalis	1	1					1	1		1	1		1	1	1	1	1	1						1	1	1	1				1	1	1	1	1	1	1	1	0
Chaetodermis pencilligerus																													1				1	0	1	0	0	0	1
Oxymonacanthus longirostris	1	1		1																													1	1	1	1	1	0	0
Paraluteres prionurus	1	1					1							1												1						1	1	1	1	1	1	1	0
Paramonacanthus curtorhynchos																						1											1	0	1	0	0	1	0
Pervagor janthinosoma		1						1					1																				1	1	1	1	1	0	0
Pervagor melanocephalus	1	1																								1							1	1	1	1	0	1	0
Pseudalutarius nasicornis									1														1										1	0	1	0	1	1	0
Pseudomonacanthus macrurus	1																					1											1	1	1	0	0	1	0
Ostraciidae (5 spp.)																																	0	0	0	0	0	0	0
Lactoria diaphanus											1																						1	0	1	0	1	0	0
Lactoria fornasini																																1	1	0	1	0	0	0	0
Ostracion cubicus	1	1		1	1			1	1	1	1	1	1	1	1	1	1	1	1							1							1	1	1	1	1	1	0
Ostracion meleagris	1	1				1		1	1	1	1				1	1	1				1					1							1	1	1	1	1	1	0
Ostracion solorensis																		1							1							1	1	0	1	0	0	1	0
Tetraodontidae (15 spp.)																																	0	0	0	0	0	0	0
Arothron caeruleopunctatus		1																															0	1	1	1	0	0	0
Arothron hispidus	1	1						1	1																								1	1	1	1	1	0	0
Arothron immaculatus	1																																0	1	1	0	0	0	0
Arothron manilensis																									1								1	0	1	0	0	1	0
Arothron mappa	1												1						1					1			1						1	1	1	0	1	1	0
Arothron nigropunctatus	1	1						1	1				1	1	1	1	1	1	1	1	1			1	1	1	1						1	1	1	1	1	1	0
Arothron stellatus	1	1			1								1						1														1	1	1	1	1	1	0
Canthigaster amboinensis		1		1		1	1							1																			1	1	1	1	1	0	0
Canthigaster axilogus		1									1																						1	1	1	1	1	0	0
Canthigaster bennetti	1	1		1	1		1																									1	1	1	1	1	1	0	0
Canthigaster compressa	1								1													1	1						1	1			1	1	1	0	1	1	1
Canthigaster epilamprus	1	1								1				1																			1	1	1	1	1	0	0

table continued on next page

Appendix 3.1. *continued*

	Previous Bali	NP Surveys	Site 1	Site 2	Site 3	Site 4	Site 5	Site 7	Site 9	Site 10	Site 11	Site 12	Site 13	Site 14	Site 15	Site 16	Site 17	Site 18	Site 19	Site 20	Site 21	Site 22	Site 23	Site 24	Site 25	Site 26	Site 28	Site 33	Site 29	Site 30	Site 31	Site 32	Present Survey	Previous Surveys	Grand Total	Nusa Penida	East Bali	North Bali	Gilimanuk
Canthigaster janthinoptera	1	1					1												1													1	1	1	1	1	1	1	0
Canthigaster papua		1			1			1																	1							1	1	1	1	1	1	1	0
Canthigaster valentini	1	1	1	1	1		1	1	1	1	1	1	1	1	1	1	1	1	1	1	1			1	1	1	1		1		1	1	1	1	1	1	1	1	1
Diodontidae (2 spp.)																																	0	0	0	0	0	0	0
Diodon hystrix		1			1					1				1			1									1			1				1	1	1	1	1	1	1
Diodon liturosus	1	1			1														1					1									1	1	1	1	1	1	0
Molidae (1 spp.)																																	0	0	0	0	0	0	0
Mola mola		1																															0	1	1	1	0	0	0
	428	573	96	162	157	91	131	187	115	183	143	117	197	190	217	220	230	246	189	99	114	42	56	191	171	248	212	2	109	85	113	139	805	641	977	573	510	622	153
74 new records for Bali	Previous Bali	NP Surveys	Site 1	Site 2	Site 3	Site 4	Site 5	Site 7	Site 9	Site 10	Site 11	Site 12	Site 13	Site 14	Site 15	Site 16	Site 17	Site 18	Site 19	Site 20	Site 21	Site 22	Site 23	Site 24	Site 25	Site 26	Site 28	Site 33	Site 29	Site 30	Site 31	Site 32	Present Survey	Previous Surveys	Grand Total	Nusa Penida	East Bali	North Bali	Gilimanuk

Chapter 4

The Status of Coral Reefs in Bali

Muhammad Erdi Lazuardi, I Ketut Sudiarta, I Made Jaya Ratha, Eghbert Elvan Ampou, Suciadi Catur Nugroho and Putu Liza Mustika

4.1 INTRODUCTION

Live coral coverage is important to support reef fish communities, provide renewable resources (e.g., seafood, seaweed, medicines), protect shorelines and attract domestic and international divers to foster the local economy (Chabanet et al. 1997; Cesar 2000; Musa 2002). The overall coral coverage of Bali is an indicator of coral health and is important for future management actions (Hill & Wilkinson 2004). Healthy and diverse coral coverage also contributes to visitor satisfaction (Musa 2002), which may eventually be linked to repeat visitation, visitors promoting the tourist package to others, and increased local income (see Mustika 2011).

Cesar (2000) has listed several classical threats to coral reefs, for example, poison fishing, blast fishing, over fishing, coral mining, sedimentation, urban pollution and waste, coral bleaching and unsustainable tourism. All of these threats are currently present in Bali. Accordingly, a circum-Bali snapshot of coral coverage should give an understanding of the overall health status of coral reefs in Bali. This chapter includes information on substrate coverage, hard coral genus composition and the Mortality Index of the coral reef ecosystem surveyed.

4.2 METHODS

4.2.1 Time

The Bali Marine Rapid Assessment Program (MRAP) was conducted from 29 April to 11 May 2011. Coral reef data were taken from 27 out of 32 sampled sites.

4.2.2 Survey location

Survey locations were potential Marine Protected Area (MPA) sites suggested by various stakeholders. The locations were also chosen based on ecosystem representativity. Sites within the locations were also chosen based on representativity per location. Table 4.1 and Figure 4.1 present the survey locations and sites.

4.2.3 Survey method

A modified point intercept transect method was used for coral reef data collection (English et al. 1997), utilizing transect lines of 2 × 50m parallel to the coastline at two depths (5–7m and 10–14m). Survey points were made every 0.5m per transect. Benthic substrates observed were hard corals (to genus level), soft corals, dead corals, rubbles, other fauna and abiotic components.

Table 4.1. Survey sites and locations during the Bali MRAP 2011

	Site	Location	Site #	Geographical coordinates	
				Longitude	Latitude
1	Kutuh	Nusa Dua	4	115.20685	-8.84418
2	Nusa Dua	Nusa Dua	5	115.23918	-8.79997
3	Melia Bali	Nusa Dua	6	115.23660	-8.79276
4	Terora	Nusa Dua	1	115.22960	-8.77044
5	Sanur Channel	Sanur	3	115.27136	-8.71027
6	Glady Willis	Sanur	2	115.26820	-8.68409
7	Tanjung Jepun	Padangbai	9	115.50976	-8.51941
8	Gili Batutiga/Mimpang	Candidasa	7	115.57488	-8.52524
9	Gili Tepekong	Candidasa	10	115.58612	-8.53141
10	Gili Biaha	Candidasa	11	115.61290	-8.50379
11	Seraya	Seraya	12	115.68918	-8.43350
12	Gili Selang	Seraya	13	115.71062	-8.39677
13	Bunutan	Amed	15	115.67892	-8.34503
14	Jemeluk	Amed	16	115.66142	-8.33737
15	Kepah	Amed	17	115.65391	-8.33384
16	Tukad Abu	Tulamben	18	115.61071	-8.29312
17	Tulamben Drop off	Tulamben	19	115.59726	-8.27829
18	Geretek	Tejakula	20	115.41447	-8.15106
19	Penuktukan	Tejakula	21	115.39587	-8.13868
20	Takad Pemuteran	Pemuteran	24	114.66682	-8.12953
21	Sumberkima	Pemuteran	25	114.60703	-8.11196
22	Anchor Wreck	P. Menjangan	26	114.50653	-8.09171
23	Coral Garden	P. Menjangan	27	114.51936	-8.09158
24	Post 2	P. Menjangan	28	114.52685	-8.09687
25	Pulau Burung	Teluk Gilimanuk	30	114.45142	-8.16267
26	Klatakan Barat	Melaya	31	114.45432	-8.23189
27	Klatakan Timur	Melaya	32	114.45653	-8.23306

4.2.4 Data Analysis

The output of data collection is: the percentage of live coral coverage and composition of hard coral genera; percentage of algae coverage, other biota, rubble, abiotic components; and a Mortality Index.

Live coral coverage was calcuated based on the following formula:

$$L = \frac{\sum Li}{N} \times 100\%$$

Remarks:
L n=Percentage of sightings
Li n=The amount of sighting i
N n=The amount of sampling sites per 100 m

Live coral (hard and soft) percentage was based on the categories of Gomez & Yap (1988):

Bad : 0–24.9%
Medium : 25–49.9%
Good : 50–74.9%
Excellent : 75–100%

Mortality Index is an index for estimating the health or condition of a coral reef ecosystem (Gomez & Yap 1988). The formula is as follows:

$$MI = \frac{\text{Percentage of dead corals}}{\text{Percentage of live corals + Percentage of dead corals}}$$

Remarks : MI = Mortality Index

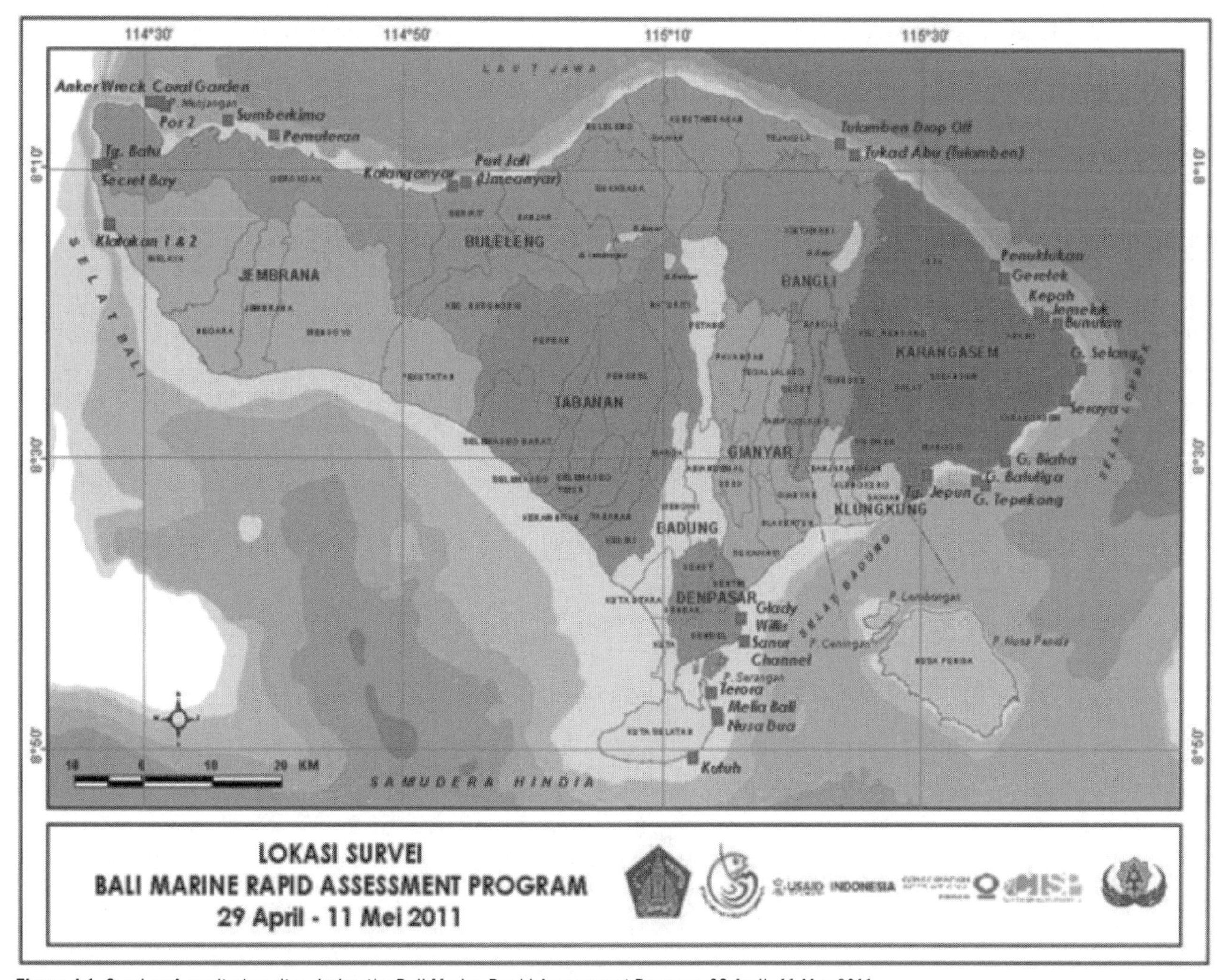

Figure 4.1. Coral reef monitoring sites during the Bali Marine Rapid Assessment Program, 29 April–11 May 2011

Table 4.2. Codes and categories for benthic life forms

Categories		Code
Hard Coral		
Acropora	Branching	ACB
	Digitate	ACD
	Encrusting	ACE
	Submassive	ACS
	Tabular	ACT
Non *Acropora*	Genus names	-
Dead Coral		DC
Dead Coral with Algae		DCA
Other Fauna		
Soft Coral		SC
Sponges		SP
Zoanthids		ZO
Others		OT

Categories		Code
Algae	Algal Assemblage	AA
	Coralline Algae	CA
	Halimeda	HA
	Macro Algae	MA
	Turf Algae	TA
Abiotic	Sand	S
	Rubble	R
	Silt	SI
	Rock	RC

Source: English et al., 1997

The Mortality Index ranges between 0–1. A near 0 Mortality Index indicates that the coral reef ecosystem is healthy with low mortality. On the other hand, a near 1 Mortality Index indicates an unhealthy coral reef ecosystem with high mortality.

4.3. RESULTS AND DISCUSSION

4.3.1 Percentage of substrate coverage

Benthic substrates were grouped into hard coral, soft coral, algae, other biota (i.e. sponge, zoanthid and other benthic biota), dead coral (i.e. dead coral and algae-covered dead coral), rubble and other abiotic components (sand, rock and mud).

4.3.2 Percentage of hard coral cover

Hard cover percentage at the 5–7m depth ranged from 21.5% to 68.0%. Site 26 (Anchor Wreck, Menjangan Island) had the highest hard coral cover, while Site 32 (East Klatakan, Melaya) had the lowest hard coral cover. The average hard coral cover at the 5–7m depth was 45.3%. On average, hard coral still dominated other substrates, for example abiotic (17.3%) and rubble (11.3%).

Hard coral cover at 10–14m ranged between 11.0% and 76.0%. Site 10 (Gili Tepekong) had the highest hard coral cover, while Site 4 (Kutuh) had the lowest hard coral cover. The average hard coral cover at this depth was 32.8%. On average, hard coral still dominated other substrates, for example abiotic (14.9%) and rubble (13.6%). Overall, the average of hard coral cover in Bali was 38.2%, ranging between 11.0 and 76.0%.

4.3.3 Coverage of other substrates

Soft corals were observed to dominate Sites 4, 5, 6, and 12 with the average percentage of cover ranging between 57.5 and 62.0%. On the other hand, abiotic substrates dominated Sites 2, 15, 18, 24, and 32 with the average percentage of cover ranging between 36.3 and 48.0%.

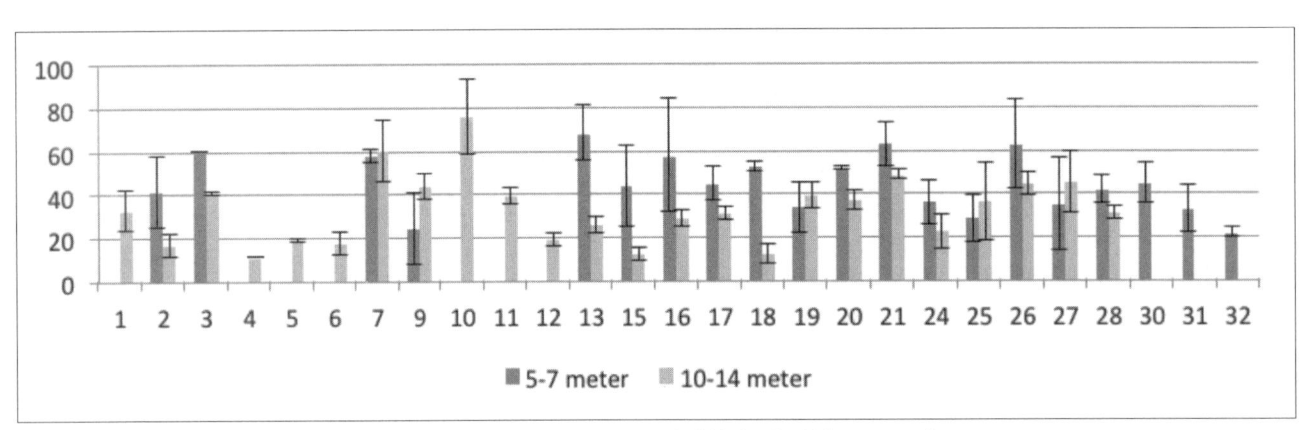

Figure 4.2. Hard coral coverage at 5–7m and 10–14m on survey sites during the Bali Marine Rapid Assessment Program

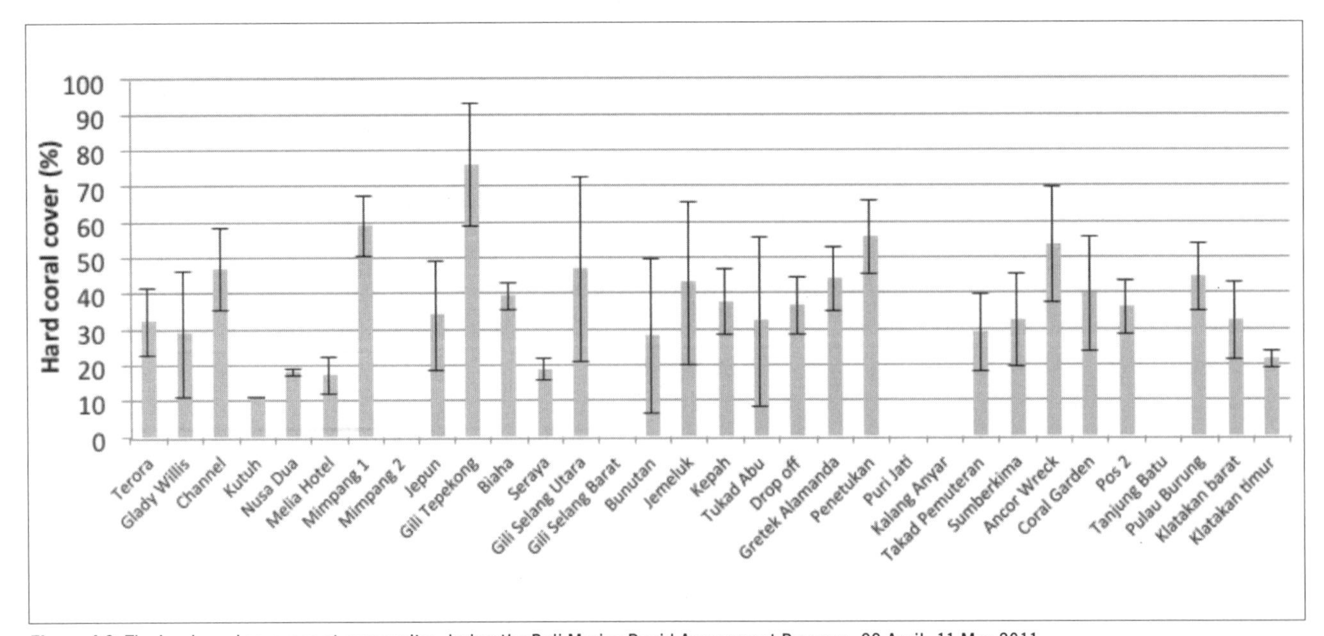

Figure 4.3. The hard coral coverage at survey sites during the Bali Marine Rapid Assessment Program, 29 April–11 May 2011

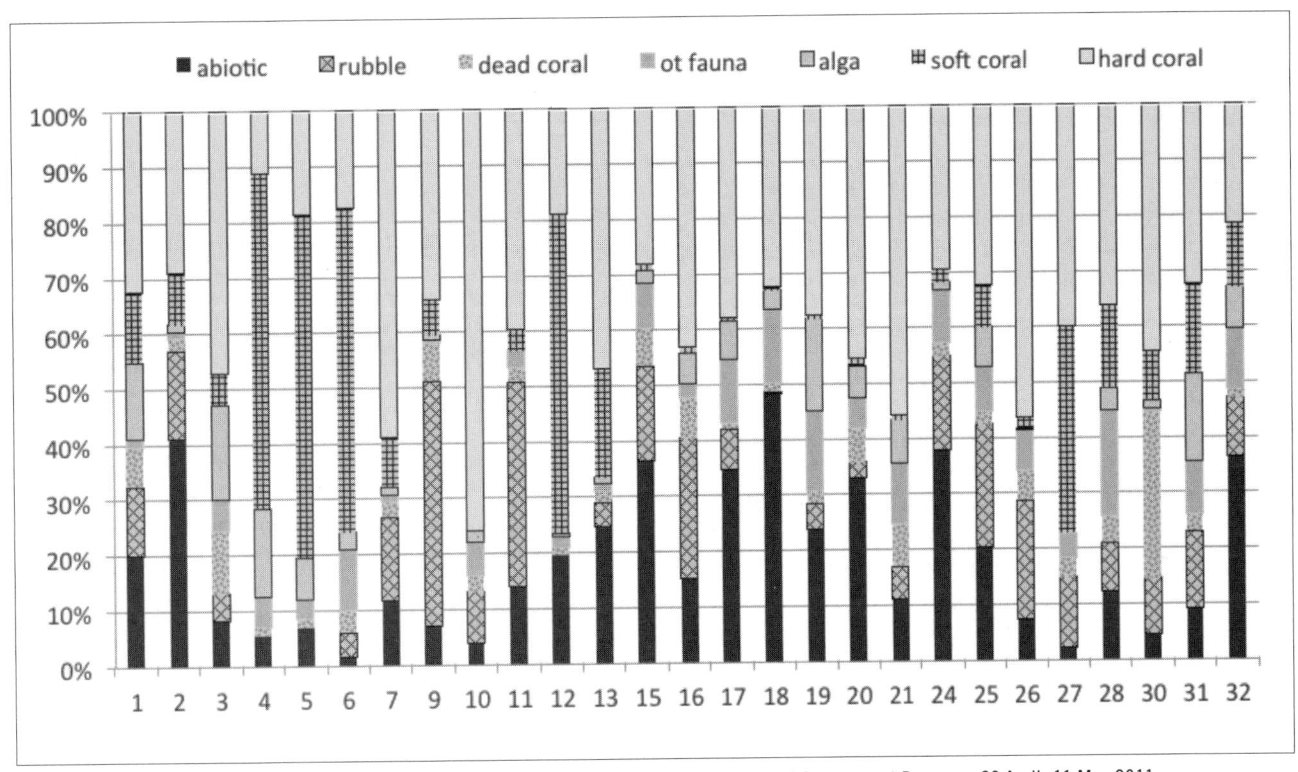

Figure 4.4. Average coverage of benthic substrates at survey sites during the Bali Marine Rapid Assessment Program, 29 April–11 May 2011

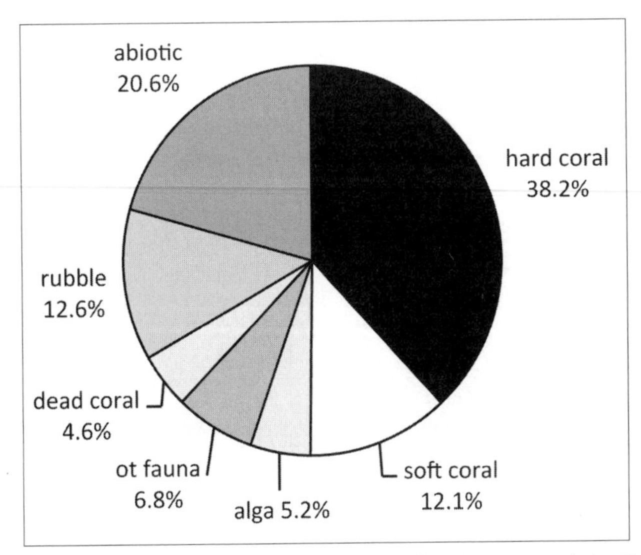

Figure 4.5. The average composition of total substrate coverage during the Bali Marine Rapid Assessment Program, 29 April–11 May 2011

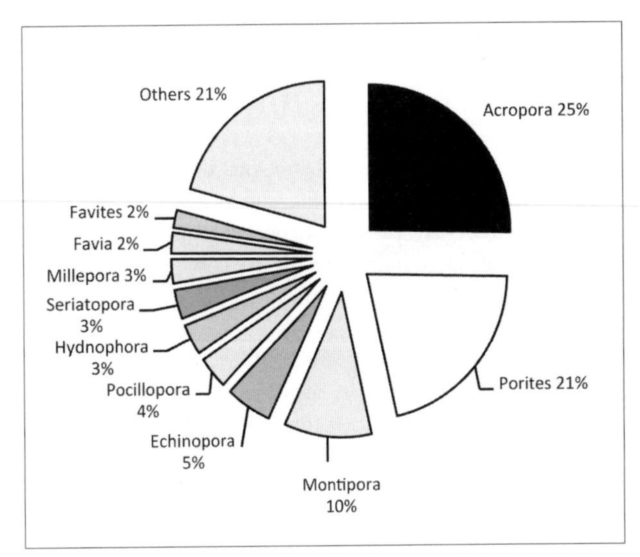

Figure 4.6. The average composition of the ten genera that dominated hard corals found during the Bali Marine Rapid Assessment Program, 29 April–11 May 2011

Sites with the highest rubble coverage were Site 9 (Jepun, on average 44.3 %), Site 11 (Biaha, on average 37.0 %), and Site 16 (Jemeluk, on average 25.3 %). Rubble coverage at other sites ranged between 0 and 22.3 %.

The highest dead coral cover (i.e. dead coral + dead coral with algae) was at Site 30 (Burung Island, Gilimanuk, on average 30.0 %). Other sites had dead coral cover ranging from 1.0–11.3 %. The average algal coverage was between 0 and 17.0 %; other fauna coverage was between 0.5 and 19.0 %.

Overall, hard corals dominated the substrates at the 5–7m and 10–14m depths (average coverage 38.2 %), outranking abiotic components (20.6 %), rubbles (12.6 %), soft corals (12.1 %), other biota (6.8 %), algae (5.2 %) and dead coral (4.6 %).

4.3.4 Live coral cover (Hard Coral + Soft Coral)

The 5–7 m water depth

The percentage of live corals (hard coral + soft coral) at the 5–7m water depth ranged between 31.5 and 85.0 %. Site 27 (Coral Garden, P. Menjangan) had the highest percentage, while Site 25 (Sumber Kima) had the lowest percentage. The overall status of live coral cover at the 5–7m depth in Bali is as follows:

Data unavailable (7 sites) : Sites 1, 4, 5, 6, 10, 11, and 12
Bad : –
Medium (9 sites) : Sites 2, 9, 15, 17, 19, 24, 25, 31, and 32
Good (9 sites) : Sites 3, 7, 16, 18, 20, 21, 26, 28, and 30
Excellent (2 sites) : Sites 13 and 27

The coral cover at the 5–7m depth was good with a live coral coverage of 54.2 %.

The 10–14 m water depth

The live coral cover (hard coral and soft coral) at 10–14m ranged between 12.0 and 80.5 %. The highest coverage was at Site 5 (Nusa Dua), the lowest coverage was at Site 18 (Tukad Abu). The overall status for live coral cover at the 10–14m depth in Bali is as follows:

Data unavailable (3 sites) : Sites 30, 31, and 32
Bad (3 sites) : Sites 15, 18, and 24
Medium (14 sites) : Sites 1, 2, 3, 9, 11, 13, 16, 17, 19, 20, 21, 25, 26, and 28
Good (3 sites) : Sites 4, 7, and 27
Excellent (4 sites) : Sites 5, 6, 10, and 12

The coral cover at 10–14m was medium with live coral coverage of 47.7 %.

Overall, live corals at the 5–7m and 10–14m bathymetries were good with average coral cover of 50.4 %.

4.3.5 The composition of hard coral genera

The observed hard coral genera consisted of reef-building corals (zooxanthellae) and non reef-building corals (non zooxanthellae). The point intercept transect method recorded 54 hard coral genera with 0.01–9.67 % coral cover per site (average 38.16 %). *Acropora* was the dominant genus with 9.67 % average coverage, outranking *Porites* (8.12 %) and *Montipora* (3.92 %). These three genera were dominant across all sites.

Table 4.3. The average status of live corals and Mortality Indexes on survey sites during the Bali Marine Rapid Assesment Program, 29 April–11 May 2011

Site no.	Site	Depth	Genus	% cover
3	Sanur Channel	5–7 m	Acropora (*branching*)	31.00 %
7	Batu Tiga/ Mimpang	5–7 m	Acropora (*branching*)	56.00 %
7	Batu Tiga/ Mimpang	10–14 m	Acropora (*branching*)	46.00 %
7	Batu Tiga/ Mimpang	10–14 m	Acropora (*branching*)	58.00 %
9	Tj. Jepun	5–7 m	Acropora (*branching*)	29.00 %
9	Tj. Jepun	10–14 m	Acropora (*branching*)	35.00 %
10	Gili Tepekong	10–14 m	Echinopora	26.00 %
10	Gili Tepekong	10–14 m	Echinopora	74.00 %
13	Gili Selang	5–7 m	Acropora (*branching*)	50.00 %
13	Gili Selang	5–7 m	Acropora (*branching*)	47.00 %
15	Bunutan	5–7 m	Porites	32.00 %
16	Jemeluk	5–7 m	Acropora (*submassive*)	24.00 %
16	Jemeluk	5–7 m	Porites	23.00 %
19	Tulamben Drop off	10–14 m	Montipora	27.00 %
25	Sumberkima	5–7 m	Acropora (*branching*)	22.00 %
26	Anchor Wreck	5–7 m	Porites (*branching*)	45.00 %
26	Anchor Wreck	5–7 m	Porites (*branching*)	43.00 %
26	Anchor Wreck	10–14 m	Porites (*branching*)	22.00 %
27	Coral Garden	5–7 m	Porites (*branching*)	26.00 %
27	Coral Garden	10–14 m	Porites (*branching*)	23.00 %
30	Pulau Burung	5–7 m	Seriatopora	51.00 %

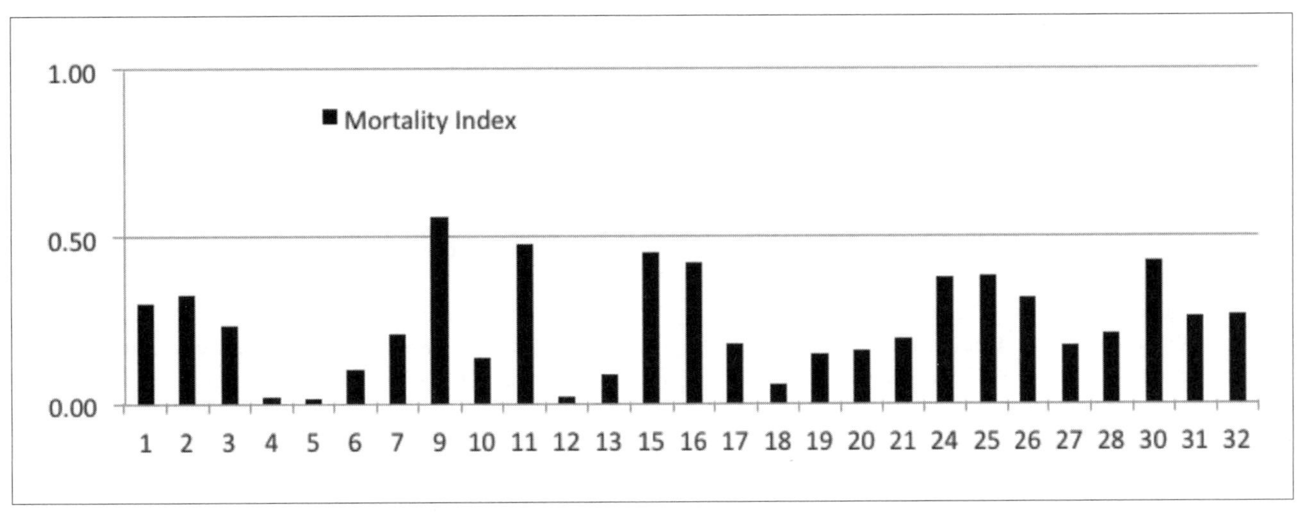

Figure 4.7. The coral reef Mortality Indexes at survey sites during the Bali Marine Rapid Assessment Program, 29 April–11 May 2011

Table 4.4. The average status of live corals and Mortality Indexes on survey sites during the Bali Marine Rapid Assesment Program, 29 April–11 May 2011

Site #	Site name	Location	Live coral status	Mortality Index
1	Terora	Nusa Dua	Medium	0.30
2	Glady Willis	Sanur	Medium	0.32
3	Channel	Sanur	Good	0.23
4	Kutuh	Uluwatu	Good	0.02
5	Nusa Dua	Nusa Dua	Excellent	0.02
6	Melia Hotel	Nusa Dua	Excellent	0.10
7	Batu Tiga/Mimpang	Candi Dasa	Good	0.21
9	Jepun	Padang Bai	Medium	0.56
10	Gili Tepekong	Candi Dasa	Excellent	0.14
11	Biaha	Candi Dasa	Medium	0.48
12	Seraya	Seraya	Excellent	0.02
13	Gili Selang Utara	Gili Selang	Good	0.09
15	Bunutan	Amed	Medium	0.45
16	Jemeluk	Amed	Medium	0.42
17	Kepah	Amed	Medium	0.18
18	Tukad Abu	Tulamben	Medium	0.06
19	Drop off	Tulamben	Medium	0.15
20	Gretek Alamanda	Tejakula	Medium	0.16
21	Penuktukan	Tejakula	Good	0.20
24	Takad Pemuteran	Pemuteran	Medium	0.38
25	Sumberkima	Pemuteran	Medium	0.38
26	Anchor Wreck	P. Menjangan	Good	0.31
27	Coral Garden	P. Menjangan	Excellent	0.17
28	Pos 2	P. Menjangan	Good	0.21
30	Pulau Burung	Gilimanuk	Good	0.43
31	Klatakan Barat	Melaya	Medium	0.26
32	Klatakan Timur	Melaya	Medium	0.27

The entire group of hard coral genera consisted of *Acropora* (25.3%), *Porites* (21.3%) and *Montipora* (10.3%). Deconstructed, the *Acropora* comprised branching *Acropora* (75%), tabulate *Acropora* (15%), submassive *Acropora* (7%), encrusting *Acropora* (2%) and digitate *Acropora* (1%). Figure 4.6 describes the ten major genera of hard corals found during the survey.

Acropora dominated the hard coral cover at Sanur Channel, Batu Mimpang, Tanjung Jepun, Gili Selang, and Sumberkima. The 10–14m bathymetry of Gili Tepekong was dominated by *Echinopora*. *Porites* dominated Bunutan, Jemeluk, Anchor Wreck and Coral Garden. Slightly below *Porites*, submassive *Acropora* were found in Jemeluk at the 5–7 m depth. *Montipora* dominated the 5–7m depth of Tulamben Drop off. *Seriatopora* dominated the Burung Island of Gilimanuk.

4.3.6 Mortality Index

The Mortality Index is a measure of coral mortality or the status of coral health. The Mortality Index of all survey sites in Bali ranged between 0.02 and 0.56.

Sites 4 (Kutuh) and 5 (Nusa Dua) had the lowest Mortality Index, which indicated that both sites had the lowest coral mortality and the highest coral health compared to other sites. Site 5 had excellent coral cover. However, both sites were dominated by soft corals. The highest Mortality Index was found in Site 9 (Jepun), which indicated an unhealthy coral reef ecosystem with high coral mortality.

The total average Mortality Index for survey sites in Bali was 0.24. Based on the above histogram, the reefs in Bali were relatively healthy with low mortality.

4.4 CONCLUSION

The data collected from the 27 sites showed that Bali's coral reefs were in good condition with an average overall coral cover of 52.3%. The average hard coral cover was 38.2%. The average Mortality Index was 0.24. These statistics indicated relatively healthy reefs with low mortality.

Based on the percentage of live coral cover (hard and soft corals), the best coral reefs at a 5–7m bathymetry were found at Site 27 (Coral Garden, P. Menjangan) while the worst was found at Site 25 (Sumber Kima). The best live coral cover at a 10–14m bathymetry was found at Site 5 (Nusa Dua), the worst was found at Site 18 (Tukad Abu). When translated into the categories of Gomez and Yap (1998), the coral reefs at survey sites in Bali were on average good (ranging from medium to excellent). The best live coral cover was found at Site 5 (Nusa Dua) with 80.5% coverage (i.e. excellent). The worst live coral cover was found at Site 29 (Bunutan) with 29.0% coverage (i.e. medium). Overall, the coral coverage at the 5–7m depth was better than that at the 10–14m depth.

The dominant hard coral genera were *Acropora*, followed by *Porites* and *Montipora*. *Acropora* was dominated by the branching form. A total of 54 hard coral genera were recorded during the survey.

REFERENCE

Cesar, H. S. J. 2000, 'Coral Reefs: Their Functions, Threats and Economic Value', in *Collected essays on the economics of coral reefs*, ed. H. S. J. Cesar, CORDIO, Kalmar.

Chabanet, P., Ralambondrainy, H., Amanieu, M., Faure, G. & Galzin, R. 1997, 'Relationships between coral reef substrata and fish', *Coral Reefs*, vol. 16, no. 2, pp. 93–102.

English, S., Wilkinson, C. & Baker, V. 1997, *Survey Manual for Tropical Marine Resources (2nd Edition)*, Australian Institute of Marine Science, Townsville.

Gomez, E. D. & Yap, H. T. 1988, 'Monitoring Reef Conditions', in *Coral Reef Management Handbook*, eds R. A. Kenchington & B. E. T. Hudson, Unesco Regional Office for Science and Technology for South-East Asia, Jakarta.

Hill, J. & Wilkinson, C. 2004, *Methods for Ecological Monitoring of Coral Reefs*, Australian Institute of Marine Science, Townsville.

Musa, G. 2002, 'Sipadan: a SCUBA-diving paradise: an analysis of tourism impact, diver satisfaction and tourism management', *Tourism Geographies*, vol. 4, no. 2, pp. 195–209.

Mustika, P. L. K. 2011, 'Towards Sustainable Dolphin Watching Tourism in Lovina, Bali, Indonesia'. Unpublished thesis. James Cook University.

Appendix 4.1. List of hard coral genera and the average coverage per survey site during the Bali Marine Rapid Assessment Program, 29 April–11 May 2011

No.	Hard coral genera	Sightings in all transects (n=3,358 points across 88 transects)	Average coverage
1	Acropora	851	9.67%
2	Porites	715	8.12%
3	Montipora	345	3.92%
4	Echinopora	177	2.01%
5	Pocillopora	121	1.38%
6	Hydnophora	115	1.31%
7	Seriatopora	108	1.23%
8	Millepora	90	1.02%
9	Favia	77	0.88%
10	Favites	66	0.75%
11	Galaxea	63	0.72%
12	Stylophora	52	0.59%
13	Goniastrea	42	0.48%
14	Fungia	36	0.41%
15	Psammocora	35	0.40%
16	Cyphastrea	30	0.34%
17	Lobophyllia	29	0.33%
18	Pectinia	27	0.31%
19	Montastrea	26	0.30%
20	Porites s	26	0.30%
21	Symphyllia	26	0.30%
22	Oxypora	22	0.25%
23	Mycedium	21	0.24%
24	Turbinaria	21	0.24%
25	Goniopora	20	0.23%
26	Leptoseris	20	0.23%
27	Platygyra	19	0.22%
28	Echinophyllia	18	0.20%
29	Merulina	18	0.20%
30	Tubipora	18	0.20%
31	Diploastrea	16	0.18%
32	Euphyllia	15	0.17%
33	Leptoria	11	0.13%
34	Pachyseris	8	0.09%
35	Siderastrea	7	0.08%
36	Ctenactis	7	0.08%
37	Alveopora	6	0.07%
38	Herpolitha	6	0.07%
39	Pavona	6	0.07%
40	Physogyra	6	0.07%
41	Anacropora	5	0.06%
42	Caulastrea	4	0.05%
43	Halomitra	4	0.05%
44	Astreopora	3	0.03%
45	Gardineroseris	3	0.03%
46	Oulophyllia	3	0.03%
47	Podabacia	3	0.03%
48	Tubastrea	3	0.03%
49	Acanthastrea	2	0.02%
50	Sandalolitha	2	0.02%
51	Coeloseris	1	0.01%
52	Scapophyllia	1	0.01%
53	Cycloseris	1	0.01%
54	Plerogyra	1	0.01%

Chapter 5

Biodiversity and Conservation Priorities of Reef-building Corals in Bali, Indonesia

Emre Turak and Lyndon DeVantier

EXECUTIVE SUMMARY

This report describes the results of surveys of biodiversity and status of coral communities of Bali, surveyed in November 2008 (Nusa Penida area) and April–May 2011 (main island). This area forms part of the Nusa Tenggara region of the Lesser Sunda Islands, at the southern edge of the Coral Triangle (CT), earth's most diverse tropical marine province. The surveys were designed to assess biodiversity and ecological condition and identify sites of conservation priority, towards expanding and improving functionality of the Marine Protected Areas network. The surveys formed part of a collaborative project between Conservation International and partners, including the Indonesian Department of Nature Conservation (PHKA), the Indonesian Ministry of Marine Affairs and Fisheries (MMAF), the Indonesian Institute of Sciences (LIPI).

A total of 85 stations (adjacent deep and shallow areas) at 48 sites (individual GPS locations) were surveyed. Coral communities were assessed in a broad range of wave exposure, current and sea temperature regimes, and included all main habitat types: cool water rocky shores, cool water reefs with broad flats, warm water reefs with broad to narrow flats, and coral communities developed on predominantly soft substrate.

The survey area is characterized by highly localized, consistent variation in several key parameters for coral growth and reef development: current flow (ranging from ca < 1 knot to > 4 knots), temperature regime (ranging during this study from ca 23–30 C, but declining to 16 C at times) and wave energy regime (ranging from ca < 1 m – 5 m), associated respectively with exposure to the Indonesian Throughflow in Lombok Strait, localized upwellings and long-period ocean swells from the Indian Ocean.

Species richness and undescribed species

Bali host a diverse reef coral fauna, with a confirmed total of 406 reef-building (hermatypic) coral species. An additional 13 species were unconfirmed, requiring further taxonomic study. At least one species, *Euphyllia* spec. nov. is new to science, and a second, *Isopora* sp., shows significant morphological difference from described species, such that there are likely to be more than 420 hermatypic Scleractinia present, in total. Notably, several widespread species exhibit consistent local morpho-types around Bali.

Within-station (point) richness around Bali averaged 112 species (s.d. 42 spp.), ranging from a low of just two species (at site B22, a muddy non-reefal location) to a high of 181 species at B16 (Jemeluk, Amed). Other species-rich sites included Menjangan N (168 spp., site B26) and Penutukang (164 spp., site B21). These results for site and overall richness are similar to those from Bunaken National Park and Wakatobi (392 and 396 spp. respectively), higher than for Komodo and Banda Islands (342 and 301 spp.), and lower than Derewan, Raja Ampat, Teluk Cenderwasih, Fak-Fak/Kaimana and Halmahera (all with more ca 450 spp. or more).

Community structure

At site level, 5 major coral community types were identified, related to levels of exposure to waves, currents — upwelling, substrate type and geographic location. These five communities were further sub-divided into 10 main coral assemblages. Each of the five communities was characterized by a more-or-less distinctive suite of species and benthic attributes.

Coral cover

Cover of living hard corals averaged 28 %. Dead coral cover was typically low, averaging < 4 % overall, such that the overall ratio of live : dead cover of hard corals was highly positive (7 : 1), indicative of a reef tract in moderate to good condition in terms of coral cover. Areas of high soft coral cover occurred on rubble beds, likely created by earlier destructive fishing, coral predation and the localized dumping of coral down-slope during creation of algal farms. Minor evidence of recent and not-so-recent blast fishing and coral diseases were also present, the latter typically on tabular species of *Acropora*. Some localized damage from recreational diver impacts, was also apparent. A very strong stress response, in the form of cyanobacterial growth was likely linked with eutrophication and sewage seepage from coastal tourism development.

Coral injury

The above impacts notwithstanding, corals of Bali exhibited relatively low levels of recent coral injury overall, in terms of the proportion of species present that were injured and the average levels of injury to those species. This was well represented by the large monospecific stands and intact massive corals present, with little evidence of major past disturbances such as coral bleaching-related mortality triggered by elevated or depressed sea temperatures, past outbreaks of coral predators, major destructive fishing activities, diseases or other impacts. This is consistent with the high positive ratio of live:dead coral cover.

Interregional comparisons

Bali's coral faunal composition is typical of the larger region, with most species recorded being found elsewhere in the CT. The overall high similarity in species composition with other parts of Indonesia notwithstanding, several important differences were apparent among these regions in the structure of their coral communities. Bali showed closest similarity to Komodo, also in the Lesser Sunda Islands and subject to a somewhat similar environmental regime in respect of current flow and cool water upwelling. These regions showed moderate to high levels of dissimilarity from most other regions to the north, notably from the more species- and habitat-rich regions of Derewan, Sangihe-Talaud, Halmahera and the Bird's Head Seascape of West Papua.

Conservation priorities

Discovery of an undescribed species of *Euphyllia* on the E coast of Bali, and the presence of other apparently local endemic corals, notably *Acropora suharsonoi*, suggests that the region does have a degree of faunal uniqueness, possibly related to the strong current flow through Lombok Strait. In this respect, the strong ITF currents may, paradoxically, both limit and foster dispersal and recruitment in different areas respectively. Local recruitment around Nusa Penida may be restricted by the currents, which may nevertheless transport larvae further afield. Genetic, reproductive and larval settlement studies would be required to test this hypothesis. If this is the case, then the islands may require careful management of local impacts, as replenishment from outside sources may be a prolonged process.

Coral communities of Nusa Penida differ from those of the main island of Bali, and are subject to different environmental conditions and human uses, and hence may require separate management focus. Reefs of high local conservation status around Nusa Penida include those at Crystal Bay, Toya Pakeh, Sekolah Dasar and N Lembongan (Stations N3, N4, N7, N8, N14 and N17. Reefs of high conservation value around Bali were widespread along the E and N coasts, and include Jemeluk, Menjangan, Gili Tepekong, Penutukang, Bunutan, Gili Selang and Gili Mimpang (Stations B16, B26, B10, B14, B21, B15, B25, B8, B18 and B7).

All the above reefs have strong potential for development of MPAs providing sufficient logistic resources and long-term support are provided. Notably, site 26 at Menjangan already forms part of the marine protected area of Bali Barat. Reefs at Jemeluk (Amed) and around Gili Tepekong, Gili Selang and Gili Mimpang are also of very high conservation value for a number of different criteria. The Batu Tiga area in particular has strong potential for development of an MPA, given that the islands are not inhabited and the reefs are already used regularly for recreational SCUBA diving.

Additionally, the wave exposed S coast community was not thoroughly surveyed because of large ocean swell. Many of these S coast reefs are highly prized for surfing, and as such draw large numbers of tourists to Bali each year. In the latter respect, their future conservation should be considered a priority for maintaining surf tourism on the island. Further offshore, some of these areas are crucial migration corridors for cetaceans and other species.

The presence of cool water upwelling and/or strong consistent current flow in some areas (eg. Nusa Penida, E Bali, as indeed also in Komodo and other areas of Indonesia) may be particularly important in buffering the incident reefs against rising sea temperatures associated with global climate change.

There is significant potential for development of MPAs providing sufficient logistic resources and long-term support are provided. Continuing impacts, particularly from litter and other forms of pollution and from poorly regulated / managed tourism development, are of concern. In respect of establishing the MPA network, the following recommendations are made:

1. A multiple use MPA model, with different areas zoned for different levels of protection and use, is likely to be the most appropriate, given the broad range of activities that already occur on Bali's reefs. However, this model should include adequate core areas excluding extractive activities, to ensure conservation of key habitat and community types and foster replenishment.

2. As far as practicable, the MPA network should include representative and complementary areas encompassing the main coral community types (Figs. 7 and 12), and reefs of high conservation value (diversity, replenishment, rarity, Table 5.10).

3. As far as practicable, the network should include reefs subject to cool water upwelling and/or strong and consistent current flow, as a potential safeguard against increasing sea temperatures associated with global climate change over coming decades. Reefs of Nusa Penida and E Bali, particularly those under the influence of Lombok Strait, should be included in the network.

4. There are many competing uses for Bali's coastal and marine resources, and it will be challenging to achieve the right balance among different levels of protection and use. Given the overwhelming importance of ocean-based toursim (surfing, diving and swimming), particular focus should be paid to maintaining healthy and attractive reef-scapes for these activities, and hence a focus on non-destructive, non-extractive activities in core zones.

5. Once an MPA network is established, enforcement of regulations will be crucial.

6. Consideration should be given to a 'User-pays' system (eg. Bunaken National Park) whereby visitors pay a nominal fee for access. This can provide significant funds for MPA management and benefits to local communities.

In respect of litter and water quality:

7. There is a widespread issue of litter and other forms of water pollution. A number of strategies may be employed / expanded to reduce the amount / impact of plastic and other pollution by: a) encouraging traditional packaging as much as practicable; b) continuing education campaigns in local mass media and schools; c) voluntary and funded litter clean-up activities on beaches and reefs.

8. Aim to improve stream and river water quality to reduce transport of litter / pollutants to reefs by restoring riparian vegetation; and with public education campaigns re inappropriate waste disposal.

5.1 INTRODUCTION

The island of Bali, Indonesia, is situated to the west of and bordering the deep-water Lombok Strait. The larger region, collectively known as the Lesser Sunda Islands, extends from Bali in the west to Timor in the east, and has been characterized as the Lesser Sunda Ecoregion (LSE) (Green and Mous 2007). The region is located on the southern edge of the Coral Triangle (CT), renowned for its globally outstanding marine biodiversity (Figure 5.1).

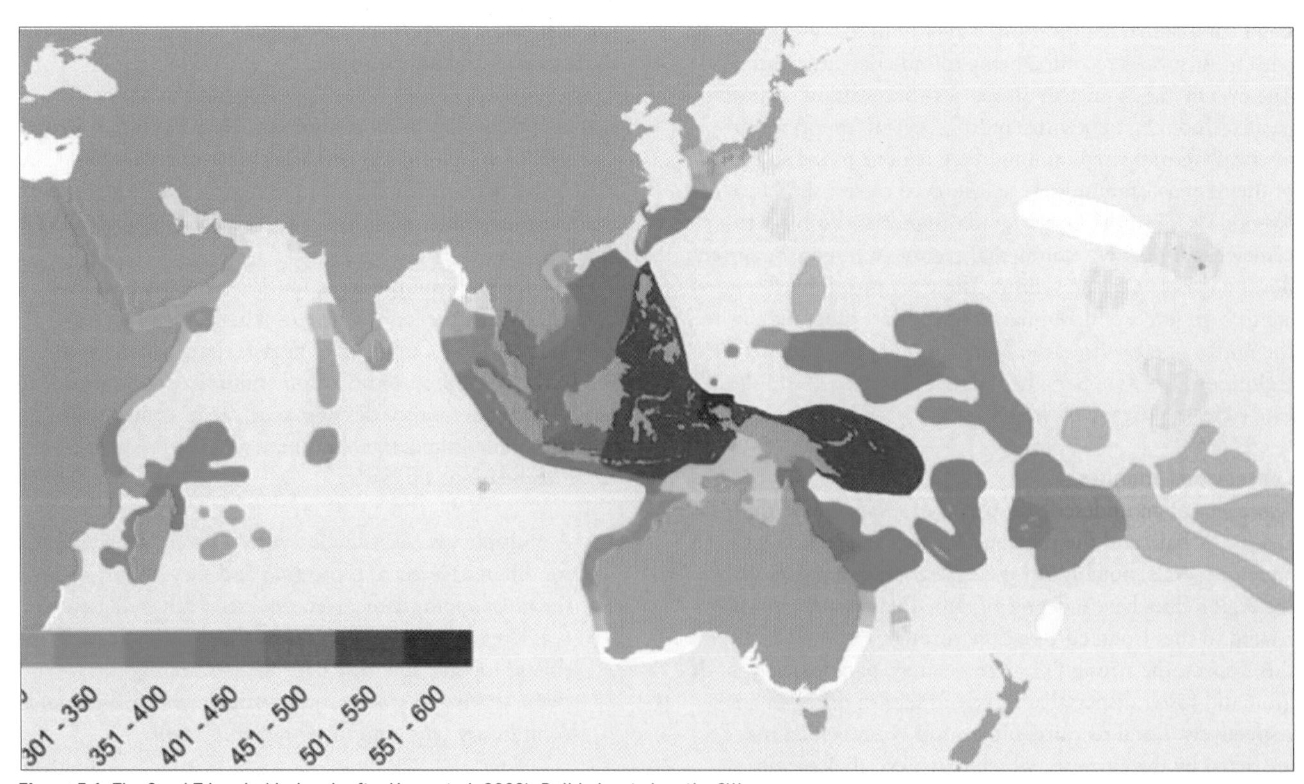

Figure 5.1. The Coral Triangle (dark red, after Veron et al. 2009). Bali is located on the SW corner.

5.1.1 Environmental Conditions and Oceanography

Bali has characteristic oceanography, tectonic–eustatic history and ecological / biological patterns. With the main Lesser Sunda island chain, Bali forms part of the north-western boundary to the Indian Ocean, and provides a major point of differentiation in several key climatological and oceanographic features.

Unlike the adjacent region to the west, which sits atop the Sunda Shelf, and regions much further east (eg. Papua) located atop the Sahul Shelf, the Lesser Sunda islands, with islands to their north, have, during the past several million years, always had deep water adjacent to their coasts. These islands have presumably played a major role as biological refugia during the Pleistocene glaciations, with significant biogeographic implications (eg. Barber et al. 2000):

> *"... there is a strong regional genetic differentiation that mirrors the separation of ocean basins during the Pleistocene low-sea-level stands, indicating that ecological connections are rare across distances as short as 300–400 km and that biogeographic history also influences contemporary connectivity between reef ecosystems."*

The Lesser Sunda Islands, including Bali, appear to be an important transition zone, with distinct faunal elements, including endemic stomatopods, fishes (M. Erdmann, G. Allen pers. comm.) and corals, and distinct coral assemblages (with relatively low coral diversity in some areas due to high wave exposure and currents).

Situated within the doldrums, Bali is sufficiently close to the equator to be unaffected directly by major tropical storms – cyclones and typhoons. There are two monsoon seasons annually, the South-east and North-west monsoons, with episodic torrential rain during the latter, from November to April. The remainder of the year is predominantly dry and hot.

The region is under the influence of the Indian Ocean Dipole (IOD). This caused anomalous upwelling, low sea surface temperatures, and low sea surface heights along the north-eastern Indian Ocean in 1997 (Abram et al. 2004, van Woesik 2004).

"Along with regional upwelling, which led to nutrient enrichment and phytoplankton blooms off the coast of Bali, there was also evidence of macroalgal blooms on the Balinese reefs. ... Coral mortality was a consequence of direct physical smothering by these macroalgae. *Acropora* and pocilloporid corals were particularly vulnerable. These corals are among the most ubiquitous, but are also the most susceptible corals in the Indian and Pacific Oceans, and are usually first to respond to any form of perturbation ... the anomalously low sea levels associated with the IOD caused direct and prolonged aerial exposure, which lead to considerable coral mortality. ... the IOD-related upwelling, independent of the wildfires, caused significant coral mortality that may have extended for at least 4000 km..." *(van Woesik 2004).*

The precise eastern extent of the influence of the 1997 IOD is not known, although the exceptionally high Chlorophyll A concentrations of September 1997 did not appear to extend eastwards beyond Bali. Strong ocean mixing typically influences both the nutrient concentrations and sea surface temperature. Sea surface productivity, as exemplified by Chlorophyll A concentration, is patchy both spatially and temporally. Waters to the south of the main island chain have higher concentrations than those to the north. Sea surface temperatures are typically cooler along the southern (Indian Ocean) coasts, particularly in the eastern and central areas (eg. Figure 5.4, May 2004). The northern coasts are usually warmer, other than in highly localized areas of upwelling.

South- and south-west facing coasts are exposed to long period ocean swell episodically exceeding 5 m height from the Indian Ocean, generated by tropical–temperate storms, many of which are thousands of km away. Bali and Lombok each host active volcanoes and the area is subject to episodic earthquakes. Tsunamis can be generated by the tectonic activity.

On its eastern shore, Bali borders Lombok Strait, with water depths greater than 1,000m in places. Lombok Strait is a major corridor of the Indonesian Throughflow (ITF), transporting Pacific Ocean water through Indonesia to the Indian Ocean. Although the main direction of water transport is from north–south, there is limited water exchange in the opposite direction. The ITF exports warm, lower salinity water from the North and central-west Pacific, providing a major water source for the north-east Indian Ocean.

> *"A net transport of nearly 20 million m3/s (Godfrey 1996) ... from the Pacific into the Indian Ocean through the Indonesian Archipelago. Originating from the Pacific, Indonesian Throughflow waters enter the Celebes Sea, move southward at velocities up to 1 m/sec (Wyrtki 1961) through the Makassar Strait, spread south and east into the Flores and Banda Seas, and ultimately exit between the Lesser Sunda Islands (Gordon & Fine 1996). Seasonally reversing east–west currents up to 75 cm/s in the Java and Flores Seas (Wyrtki 1961) further mix the surface waters."* (Barber et al. 2002).

Localized upwellings are generated by the ITF, where sea temperatures can differ by as much as 14 C within several km (ranging from 16–30 C). In addition to the effects of the ITF, local sea surface current patterns around Bali and adjacent islands are influenced by seasonal, tidal, wind and wave forcing. The long period ocean swells of the Indian Ocean impacting on the southern coastlines are likely to provide a major differentiating factor in species composition and community structure. These attenuate more or less gradually as the swell dissipates as it propagates northward between the islands.

5.1.2 Biological and biogeographical patterns and endemism

Perhaps paradoxically, the main areas of through-flow (eg. Lombok Strait) may be considered as both contributing to and restricting dispersal (see later). The local currents may prove as important as the influence of the ITF in connecting and isolating local populations.

"[While] *broad-scale oceanographic data may provide reasonable dispersal predictions…, other data may over-simplify the currents experienced by larvae that originate in near-shore environments. Eddies, stagnation zones, and local reversals of long shore currents are common in coral-reef systems, as are seasonal, tidal and weather-driven changes in current flow … These mesoscale coastal current patterns may greatly influence larval movements … and local retention … and are implicated in the formation of discrete population units … as well as in genetic structuring"* (Barber et al. 2002).

The larger Lesser Sunda Islands region hosts more than 500 species of scleractinian reef-building corals (523 species; Veron et al. 2009). Prior to the present study, a total of 12 taxonomic sites, centred on Bali and the N coast of Flores, and 104 ecological survey sites, centred on Komodo, W Lombok and W Timor – Roti, had been assessed (the former by Charlie Veron, the latter by the present authors). These locations are each dissimilar from the others, however the degree to which these locations are unique or representative of larger areas is presently unquantified.

Within the Lesser Sunda Islands, there are several sources of differentiation in respect of coral species composition and patterns in community structure, primarily caused by local – regional differences in oceanography, especially upwelling and ocean swell. Another key factor is suitability of habitat and substratum. Coastlines of Bali and adjacent islands have been formed predominantly by limestone, indicating earlier periods of reef growth and deposition.

The larger region ('southern island arc') was identified as an important area of endemism within the Coral Triangle (Erdmann and Manning 1998, Wallace 1994, 1997, Allen 2007, Veron et al. 2009), hosting species that are, on present data, considered endemic or sub-endemic (occurring more or less sparsely in other areas within the CT). These discovered in the vicinity of the present study are listed below with their author and place of discovery.

Acroporidae

- *Acropora suharsonoi* Wallace, 1994 (Lombok)
- *Acropora sukarnoi* Wallace, 1997 (Bali)
- *Acropora parahemprichii* Veron, 2002 (Bali)
- *Acropora minuta* Veron, 2002 (Bali)
- *Acropora pectinatus* Veron, 2002 (Bali)

Poritidae

- *Alveopora minuta* Veron, 2002 (Bali)

Fungiidae

- *Halomitra meierae* Veron, 2002 (Bali)

Several of these species (eg. *Acropora pectinatus, Acropora sukarnoi, Alveopora minuta*) have subsequently been found elsewhere. Nevertheless, the Bali – Lombok area appears particularly interesting in respect of coral endemism.

5.1.3 Socio-economy

Bali's traditional lifestyle relied primarily on various forms of subsistence agriculture and fishing, the former flourishing on the rich volcanic soils from Bali's active volcanoes, the latter on the rich marine coastal life. This began to change rapidly in the early 1970s, with the arrival of the first wave of international travellers; and over the subsequent 40 years, surf, beach, dive and cultural tourism have all burgeoned, collectively accounting for some 80 % of the economy in the early 21st century. The following quotes are excerpted from the project background document for the present survey (M. Erdmann, CI Indonesia Marine Program):

"Bali's rich marine resources have long been an important economic asset to the island— both as a source of food security for local communities (many of whom derive a significant proportion of their animal protein needs from seafood) and also as a focus for marine tourism. Diving and snorkeling attractions such as Nusa Penida, Candi Dasa, Menjangan Island (Bali Barat National Park), and the Tulamben USS Liberty wreck have been drawing tourists into Bali's water for decades, while more recently the private marine tourism sector has expanded the menu of options to include sites like Puri Jati, Karang Anyar, and Amed. Other important economic activities in Bali's coastal zone include seaweed farming and ornamental fish collecting."

5.1.4 Development

A census of Bali's population in 2010 recorded 3,891,428 people, increasing steadily from the 2,469,930 people present in 1980, the 2,777,811 people present in 1990 and the 3,150,057 people present in 2000 (http://www.citypopulation.de/Indonesia-MU.html). The rapid increase in population and support infrastructure over the past several decades has, however, come at significant environmental cost:

"Unfortunately, rapid and largely uncoordinated development in Bali's watersheds and coastal areas, along with a lack of clear marine spatial planning for the island, has led to significant deterioration of many marine environments around Bali due to a combination of overfishing and destructive fishing, sedimentation and eutrophication from coastal development, sewage and garbage disposal at sea, and dredging/reef channel development. At this point in time, the long-term sustainability of the many important economic activities occurring in Bali's coastal zone are in question."

5.1.5 Planning for future sustainability

Given these increasing levels of threat and impact to Bali's marine and terrestrial resources, the Bali Government is presently working towards a comprehensive long-term development strategy for the island, including greatly improving spatial planning in both the terrestrial and marine areas of Bali (M. Erdmann, CI Indonesia Marine Program):

"One important part of this initiative has been the decision by the Bali Government to design and implement a comprehensive and representative network of Marine Protected Areas around the island that prioritizes sustainable and compatible economic activities (including marine tourism, aquaculture and sustainable small-scale fisheries).

To initiate the planning for this network of MPAs, … (a) multistakeholder workshop … was organized by the Marine Affairs and Fisheries Agency of Bali Province, in collaboration with the Bali Natural Resources Conservation Agency (KSDA), Warmadewa University, Udayana University, USAID, Conservation International (CI) Indonesia and local NGOs within the framework of a "Bali sea partnership". The Bali MPA Network workshop was attended by 70 participants from the provincial government, regency governments, universities, NGOs, private sectors, community groups, traditional villages forum and fishermen groups.

Importantly, the workshop participants identified 25 priority sites around Bali as the top candidates for inclusion in a network of MPAs for the island (Figure 5.2). *This list of sites included existing national/local protected areas such as Bali Barat National Park/Menjangan Island, Nusa Penida, and Tulamben, while also including a number of additional sites that currently have no formal protection."*

5.1.6 Rationale and assessment objectives

Following the 2010 workshop, CI was asked by the Bali government, via the provincial Marine Affairs and Fisheries Agency, to lead a team of local and international experts in surveying candidate MPA sites, the results to be used to provide clear recommendations on priority development sites and next steps for the design of the MPA network.

"The team has been requested to build upon the survey data compiled during the November 2008 CI-led "Marine Rapid Assessment" of the Nusa Penida reef system to provide a more comprehensive report on the biodiversity, community/assemblage structure, and current condition of coral reefs and related ecosystems around Bali. Based upon this information, the team is to provide recommendations on how to best prioritize the 25 candidate sites for inclusion in an ecologically-representative network of MPAs. This information will be used to finalize the MPA network development plan and to justify/socialize these plans to government and local community stakeholders, so must be

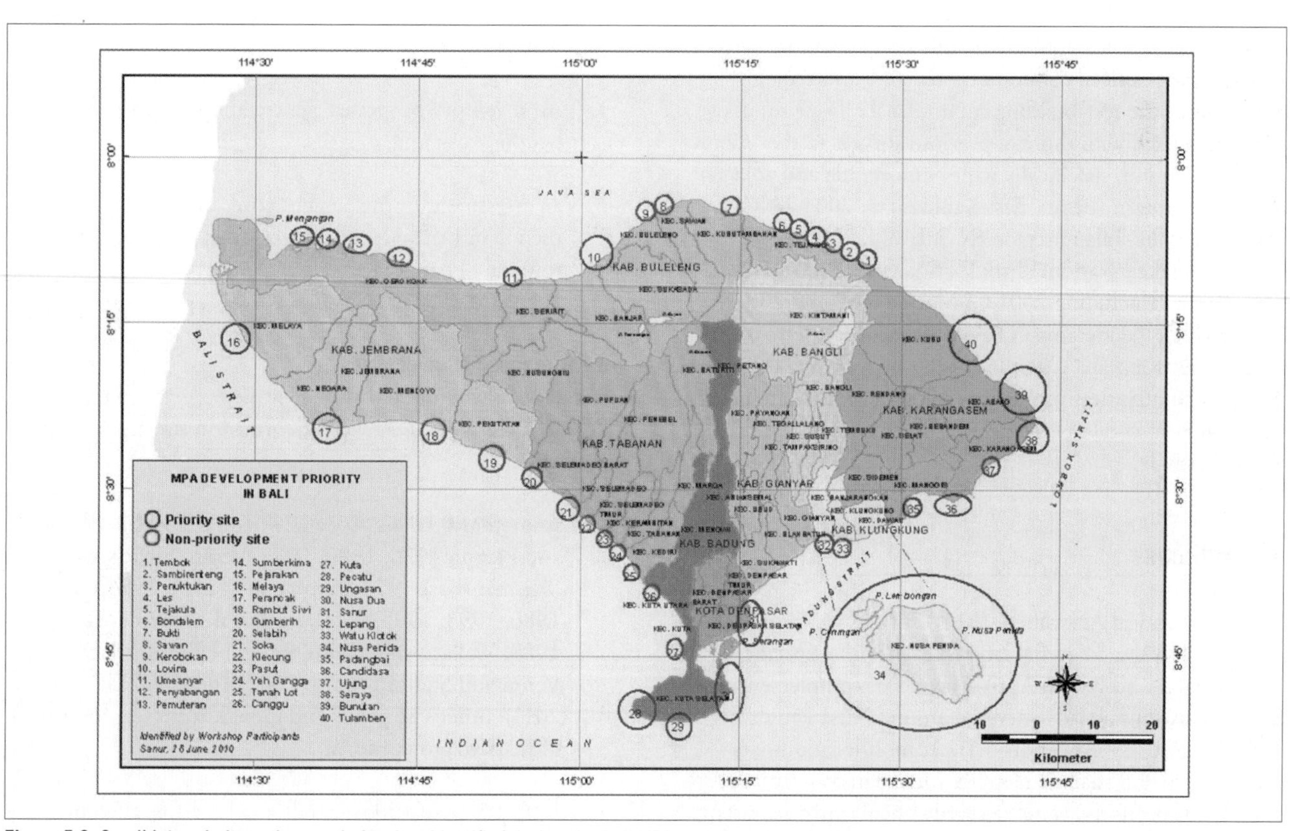

Figure 5.2. Candidate priority and non-priority sites identified during the Bali MPA workshop, June 2010.

compiled in a manner that is easily understood by a lay people audience." (M. Erdmann, CI Indonesia Marine Program 2011).

The assessment, conducted during the period of April–May 2011, has the following three primary objectives:

- Assess the current status including biodiversity, coral reef condition and conservation status/resilience of hard corals of the 25 candidate MPA sites around the island of Bali identified by the June 2010 Bali MPA Network workshop. Thorough species-level inventories of each of these groups will be compiled.

- Compile spatially-detailed data on biological features which must be taken into consideration in finalizing the Bali MPA Network design. This includes not only an analysis of any differences in reef community structure of the 25 priority sites, but also specifically identifying areas of outstanding conservation importance due to rare or endemic hard coral assemblages, reef communities exposed to frequent cold-water upwelling that may be resilient to global climate change, or other outstanding biological features.

- Taking the above into account, provide concrete recommendations to the Bali government on the next steps to be taken to finalize the design of the Bali MPA Network.

In addressing these objectives, this study documents coral species composition, community structure and ecological status of the reef-building corals of Bali. These results were compared with those of previous surveys in the "Coral Triangle" region, specifically with Nusa Penida and adjacent islands, Derewan (Berau, East Kalimantan, 2004 TNC REA), Sangihe-Talaud region (N Sulawesi, 2001 TNC REA), Bunaken National Park (N Sulawesi, 2003, IOI), Raja Ampat (including 2001 CI Marine RAP and 2002 TNC REA), Cenderawasih Bay (2006 CI Marine RAP), the FakFak/Kaimana Coastline (2006 CI Marine RAP). The goal was to quantitatively assess ecological and taxonomic similarities in coral assemblages across this section of the Coral Triangle.

5.2 METHODS

Rapid Ecological Assessment (REA) surveys were conducted using SCUBA at 31 reef locations (each with a specific GPS position) around Bali in April–May 2011, complementing the 17 locations already surveyed around Nusa Penida in 2008 (Figure 5.3, Appendix 5.1). At most locations (stations), deep and shallow reef sites (designated as station #.1 and #.2 respectively) were surveyed concurrently, representing the deeper reef slope (typically > 10m depth) and the shallow slope, reef crest and flat (typically < 10m depth), for a total of 85 stations. Deep stations were surveyed first, in accordance with safe diving practice, with the surveyor swimming initially to the maximum survey depth (usually 30–40 m), then working steadily into shallower waters. In this report, the term 'station' refers to the combined results of the two stations (depths), unless otherwise specified with the specific depth designator (station #.1 and #.2 respectively).

The method was identical to that employed during biodiversity assessments in ca. 35 other regions of Indonesia and the Indo-Pacific, providing the opportunity for detailed comparisons of species diversity, composition and community structure, and of the representativeness and complementarity of different areas in terms of their coral communities. The field and analytical methods are explained in detail elsewhere (eg. DeVantier et al. 1998).

At each station, the survey swim covered an area of approx. one ha in total. Although 'semi-quantitative', this method has proven superior to more traditional quantitative methods (transects, quadrats) in terms of biodiversity assessment, allowing for the active searching for new species records at each station, rather than being restricted to a defined quadrat area or transect line. For example, the present method has regularly returned a two- to three-fold increase in coral species records in comparison with line transects conducted concurrently at the same stations (DeVantier et al. 2004).

Two types of information were recorded on water-proof data-sheets during the ca. one and a half hour SCUBA survey swims at each station:

1. An inventory of species, genera and families of sessile benthic taxa; and

2. an assessment of the percent cover of the substrate by the major benthic groups and status of various environmental parameters (after Done 1982, Sheppard and Sheppard 1991).

5.2.1 Taxonomic inventories

A detailed inventory of sessile benthic taxa was compiled during each swim. Taxa were identified in situ to the following levels:

- stony (hard) corals – species wherever possible (Veron and Pichon 1976, 1980, 1982, Veron, Pichon and Wijsman-Best 1977, Veron and Wallace 1984, Veron 1986, 1993, 1995, 2000, Best et al. 1989, Hoeksema 1989, Wallace and Wolstenholme 1998, Wallace 1999, Veron and Stafford-Smith 2002, Turak and DeVantier 2011), otherwise genus and growth form (e.g. *Porites* sp. of massive growth-form).

- soft corals, zoanthids, corallimorpharians, anemones and some macro-algae – genus, family or broader taxonomic group (Allen and Steen 1995, Colin and Arneson 1995, Gosliner et al. 1996, Fabricius and Alderslade 2000);

- other sessile macro-benthos, such as sponges, ascidians and most algae – usually phylum plus growth-form (Allen and Steen 1995, Colin and Arneson 1995, Gosliner et al. 1996).

At the end of each survey swim, the inventory was reviewed, and each taxon was categorized in terms of its relative abundance in the community (Table 5.1). These ordinal ranks are similar to those long employed in vegetation analysis (Barkman et al. 1964, van der Maarel 1979, Jongman et al. 1997).

For each coral taxon present, a visual estimate of the total amount of injury (dead surface area) present on colonies at each station was made, in increments of 0.1, where 0 = no injury and 1 = all colonies dead. The approximate proportion of colonies of each taxon in each of three size classes was also estimated. The size classes were 1–10 cm diameter, 11–50 cm diameter and > 50 cm diameter (Table 5.1).

Taxonomic certainty: Despite recent advances in field identification and stabilizing of coral taxonomy (e.g. Hoeksema 1989, Veron 1986, Wallace 1999, Veron 2000, Veron and Stafford-Smith 2002), substantial taxonomic uncertainty and disagreement among different workers remains (Fukami et al. 2008). This is particularly so in the families Acroporidae and Fungiidae, with different workers each providing different taxonomic classifications and synonymies for various corals (see e.g. Hoeksema 1989, Sheppard and Sheppard 1991, Wallace 1999, Veron 2000). The analyses herein rely on our synthesis and interpretation of these revisions and with particular reliance on the species distribution maps of Veron (2000), currently being updated in the biogeographic database Coral Geographic (www.coralreefresearch.org).

Extensive use of digital underwater photography and a limited collection of specimens of taxonomically difficult reef-building coral species were made, in collaboration with Indonesian colleagues, notably Mr. Erdi Lazuardes of CI Indonesia, and the Indonesian Institute of Sciences, to aid in confirmation of field identifications.

Small samples, usually < 30 cm on longest axis, were removed from taxonomically-difficult corals in situ, leaving the majority of the sampled colonies intact. Living tissue was removed from the specimens by bleaching with household bleach. Many of these specimens were identified, using the above reference materials, during and following the survey, and have been deposited for short-term storage at the CI Office, Bali.

5.2.2 Benthic cover and reef development

At completion of each survey swim, six ecological and six substratum attributes were assigned to 1 of 6 standard categories (Table 5.2), based on an assessment integrated over the length and depth range of the swim (after Done 1982, Miller & De'ath 1995). Because the cover estimates apply for the area and depth range over which each survey swim was conducted (eg. ca 40 – 9m depth; 8 – 1m depth respectively), these may not correspond precisely with line transect estimates made at a single depth or set of depths (ed: see Chapter 3).

The stations were classified into one of four categories based on the amount of biogenic reef development (after Hopley 1982, Hopley et al. 1989, Sheppard & Sheppard 1991):

1. Coral communities developed directly on non-biogenic rock, sand or rubble;
2. Incipient reefs, with some calcium carbonate accretion but no reef flat;
3. Reefs with moderate flats (< 50m wide); and
4. Reefs with extensive flats (> 50m wide).

Table 5.1. Categories of relative abundance, injury and sizes (maximum diameter) of each benthic taxon in the biological inventories.

Rank	Relative abundance	Injury	Size frequency distribution
0	absent	0–1 in increments of 0.1	proportion of corals in each of 3 size classes: 1) 1–10 cm 2) 11–50 cm 3) > 50 cm
1	rare		
2	uncommon		
3	common		
4	abundant		
5	dominant		

Table 5.2. Categories of benthic attributes

Attribute		Ranks used in calculating Replenishment index CI	
Ecological	Physical	% cover	Rank
Hard coral	Hard substrate	0	0
Dead standing coral	Continuous pavement	1–10 %	1
Soft coral	Large blocks (diam. > 1 m)	11–30 %	2
Coralline algae	Small blocks (diam. < 1 m)	31–50 %	3
Turf algae	Rubble	51– 75 %	4
Macro-alga	Sand	76–100 %	5

The stations were also classified arbitrarily on the degree of exposure to wave energy, where:

1. sheltered
2. semi-sheltered
3. semi-exposed
4. exposed

The depths of the stations (maximum and minimum in m), average angle of reef slope to the horizontal (estimated visually to the nearest 10 degrees), and underwater visibility (to the nearest m) were also recorded. The presence of any unique or outstanding biological features, such as particularly large corals or unusual community composition, and evidence of impacts, were also recorded, such as:

- sedimentation
- blast fishing
- poison fishing
- anchoring
- bleaching impact
- crown-of-thorns seastars predation
- *Drupella* snails predation
- coral diseases

All data were input to EXCEL spreadsheets for storage and analysis of summary statistics.

Rarity index

The presence of species that are rare in the study area may afford some stations greater importance than others in terms of the conservation of biodiversity of corals. An index, *RI*, to indicate the relative importance of stations based on their compliment of rare coral species was calculated for each station (after DeVantier et al. 1998):

$$RI = (\sum A_i / P_i) / 100$$

where A_i=abundance rank for the ith coral taxon at a given station (1-5, as in Table 5.2), and P_i=the proportion of all stations in which the taxon was present. This index weights species on a continuum according to their frequency in the data set, and gives highest values to stations which are least representative or most unusual faunistically (ie. with high abundance of taxa which are rare in the data set).

Coral Injury

Each coral species in the stations was assigned a score for its level of injury, from 0–1 in increments of 0.1 (from 0: no injury to any colony of that species in the station to 1: all colonies of the species were dead, see Methods above). Stations were compared for the amounts of injury to their coral communities, for the proportion of the total number of species present in each station that were injured, and the average injury to those coral species in each station.

Coral community types

Site groups defined by community type were generated by hierarchical cluster analysis using abundance ranks of all corals in the individual station inventories. The analysis used Squared Euclidean Distance as the clustering algorithm and Ward's Method as the fusion strategy to generate station groups of similar community composition and abundance. Analyses were conducted on the raw (untransformed) data. The clustering results were plotted as dendrograms to illustrate the relationships among stations in terms of levels of similarity among the different community groups. Two sets of analysis were undertaken:

i. Bali
ii. Various regional analyses of adjacent regions of the CT, including Komodo, Wakatobi, Derewan, Sangihe-Talaud, Banda Islands, Bunaken National Park, Raja Ampat, Cenderwasih Bay and Fak-Fak/Kaimana (Figure 5.4).

To facilitate accurate comparison, all datasets used in the regional analysis had been recorded during various surveys undertaken by the present authors (listed in References).

5.3 RESULTS

5.3.1 Environmental Setting

The full range of reef development occurs throughout the survey area, ranging from coral communities developed directly on non-reefal substrata, to incipient reefs with some accretion, to large sub-tidal and inter-tidal reefs with flats wider than 50 m (Table 5.3, Appendix 5.1). The coral communities were developed from low-tide level to > 60 m depth, although most coral growth occurred above 30 m depth, on slopes ranging from < 5° (reef flats) to 90° to the horizontal (vertical reef walls), the latter being uncommon (Appendix 5.1). The communities were distributed over exposure regimes from sheltered to highly exposed, related to the degree of protection provided by coastal features from oceanic swells from the Indian Ocean. Large ocean swells during the survey period precluded survey of many highly-exposed S coast locations, although Bali site 4 on the SE coast to the south of Nusa Dua is a relatively exposed site, as are Nusa Penida sites 5 and 6.

Most coral communities were developed in areas of hard reefal or non-reefal substrate (mean of 76% cover) with only small areas of sand (mean 14%), and were subject to variable levels of current flow, ranging from calm to > 2 knots, related chiefly to the influence of the ITF through Lombok Strait and tidal movements. There were usually negligible levels of sedimentation, other than at silty sites on the N coast of Bali. The typically low silt levels contributed to the relatively high water clarity, which averaged 15 m, ranging from 3 m to 30 m during the survey period (Table 5.3).

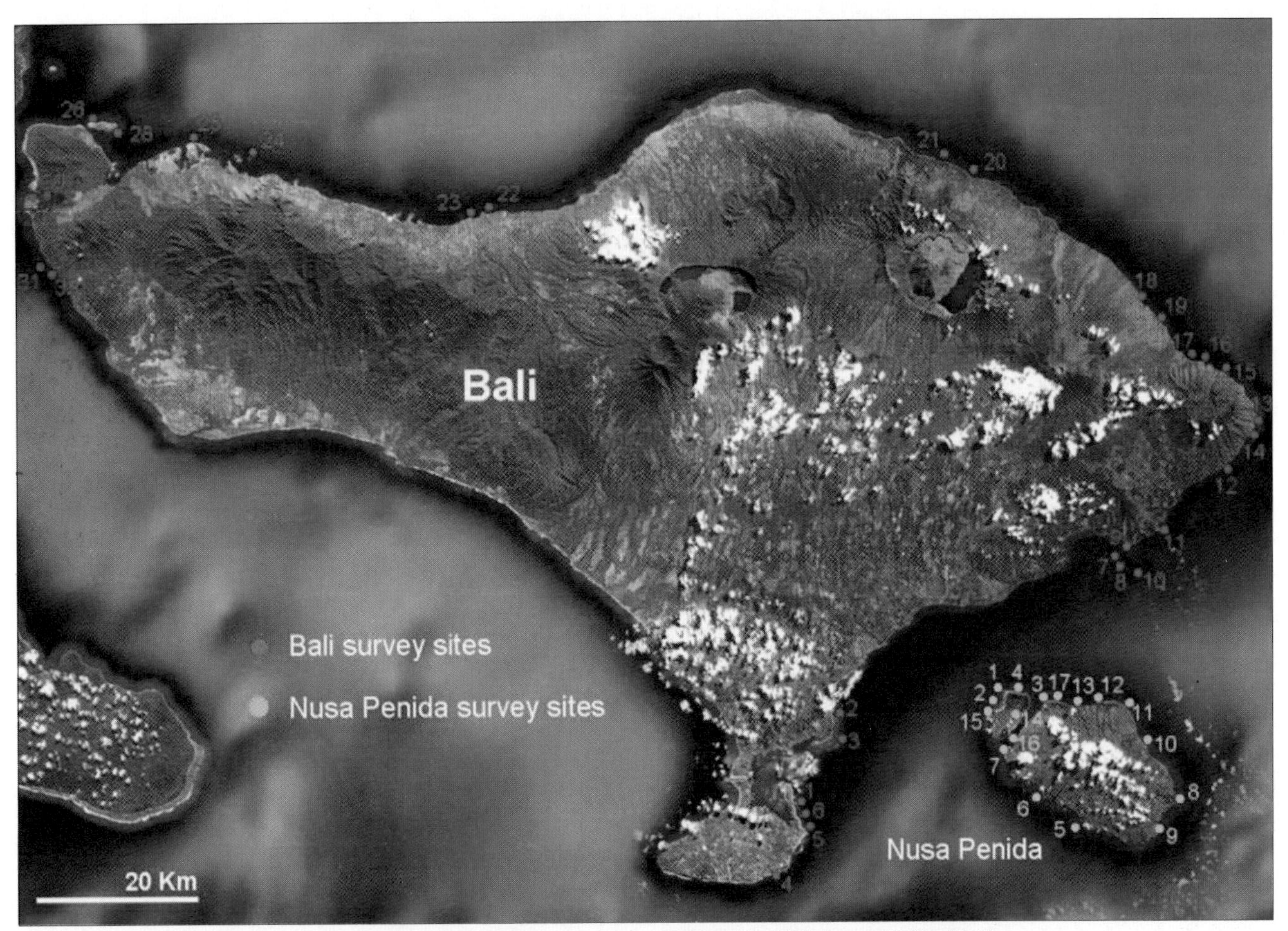

Figure 5.3. Approximate location of survey sites, Nusa Penida (17 sites, October 2008) and Bali (31 sites, April–May 2011).

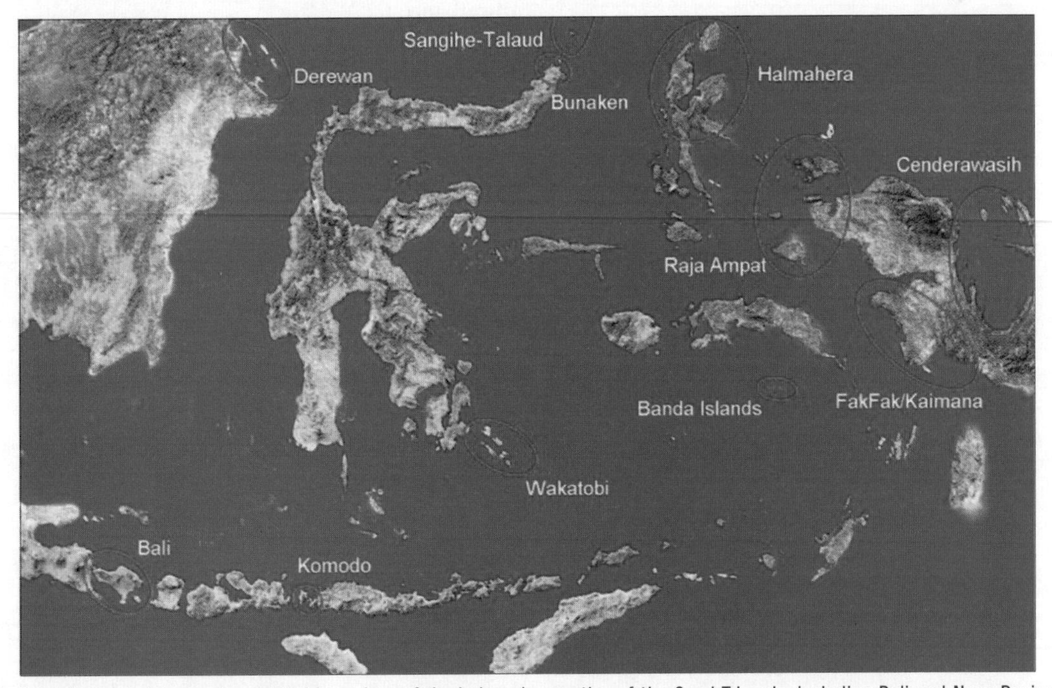

Figure 5.4. General areas of surveys conducted in regions of the Indonesian section of the Coral Triangle, including Bali and Nusa Penisa, Komodo, Banda Islands, Wakatobi, Derewan, Bunaken, Sangihe-Talaud, Halmahera, Raja Ampat, Teluk Cenderawasih and Fak-Fak/Kaimana. These survey regions are each large and support diverse reef habitats. These were each surveyed as comprehensively as practicable in the limited time available (see References for details).

Around Nusa Penida, many of the large intertidal reef flats are covered with seaweed farms. In the process of establishing and maintaining these farms virtually all live coral has been cleared from the area. While some of the excess material may have been used on land, much of it, in the form of coral rubble, including fairly large blocks, was dumped down the adjacent reef slopes. This activity has resulted into two negative impacts to the fringing reefs. First, virtually no intertidal coral community is now present, resulting in the absence, or extremely rare occurrence, of coral species limited to intertidal areas. Secondly, the loose rubble on the upper slopes is frequently moved around by the action of waves and strong currents, further damaging corals in these areas.

5.3.2 Cover of corals and other sessile benthos

Cover of living hard corals was typically moderate to high (eg. Plates 5.1–5.3), averaging 28 % (Figure 5.5), and ranging from 1–70 %. sites with high live coral cover were widespread (Appendix 5.2). Highest cover (60 % or more) occurred most commonly (but not exclusively) in shallow stations (< 10m depth), notably at Nusa Penida Stations 1.2, 3.1, 7.2 and 17.2 and Bali Stations 15.2, 26.2 and 30.2. Cover was dominated in many stations by large monospecific stands, indicative of the likely importance of asexual reproduction via fragmentation in maintaining high cover locally. Elsewhere, the presence of large intact massive corals, with little or no sign of scarring, was consistent with relatively minor long-term impact from disturbances over the past several decades.

Table 5.3. Summary statistics for environmental variables, Bali (including Nusa Penida), October 2008 and April–May 2011

Environmental variable	Mean (s.d.)	Range	Median	Mode
Reef development (rank 1–4)	2.8 (1,1)	1–4	3	4
Slope angle (degrees)	16 (15)	2–90	10	5
Exposure (rank 1–4)	2.4 (0,7)	1–4	2	2
Water Clarity (Visibility m)	15 (8)	3–30	16	20
Hard substrate (%)	76 (25)	0–100	85	90
Sand (%)	14 (18)	0–95	5	5
Water temperature (C)	28.6 (1,2)	23–30	29	29

Overall, rubble and dead corals contributed ca. 10 % cover, most of which was in the form of rubble (8 %). sites with high cover of rubble (20 % or more) included Nusa Penida Stations 7.1, 13.2, 14.1 and 15.2 and Bali Stations 7.1, 8.1, 9.1, 9.2, 11.1, 11.2, 15.1 and 16.1. The only stations with relatively high cover of standing dead corals (20 % or more) were Bali Stations 7.1, 9.1 and 9.2. Previous mortality of live corals was mostly attributable to crown-of-thorns seastar and/or *Drupella* snail predation, diseases, or algal growth from localized eutrophication. Low levels of coral diseases, particularly 'White-band' disease, were also apparent, developed primarily on tabular species of *Acropora*. There was, however, only low cover of recently killed corals (< 1 %), and the continuing minor disturbances notwithstanding, the overall ratio of live : dead cover of hard corals remained strongly positive at ca 7 : 1, indicative of a reef tract in good condition in terms of coral cover. The ratio of live hard coral cover to dead corals plus rubble was also positive at ca 5 : 2, and is consistent with these reefs supporting ca. 40 % mean live hard coral cover during periods of low disturbance.

Soft coral cover was moderate, averaging 10 % overall, and high in patches, notably on coral rubble beds. sites with high

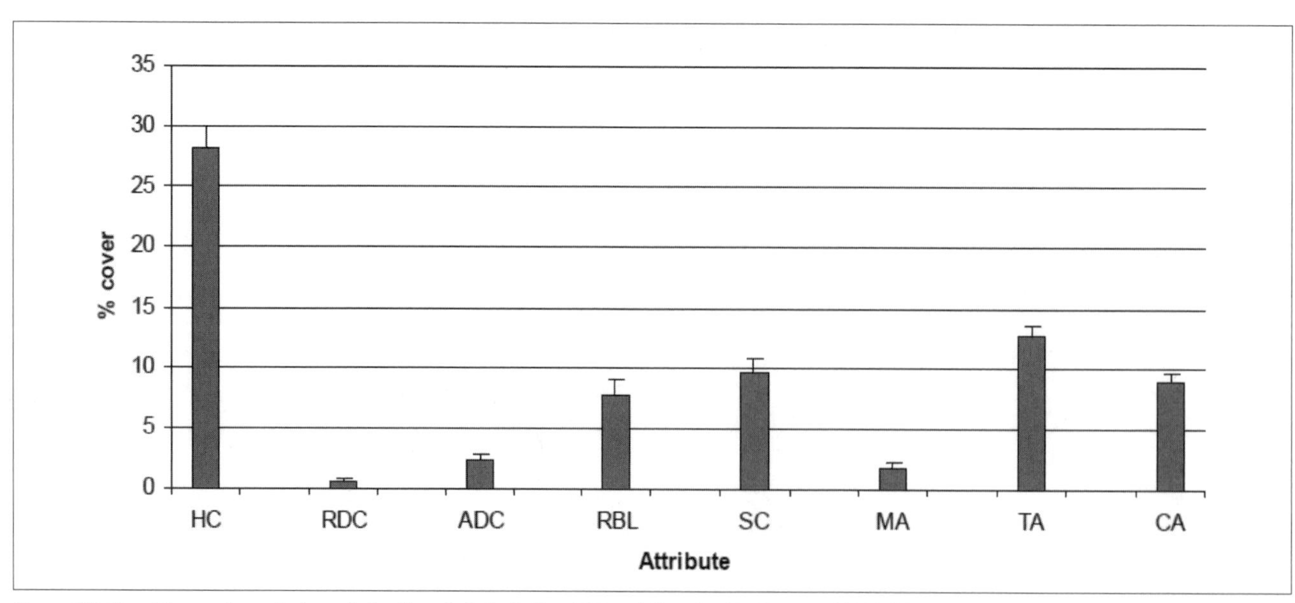

Figure 5.5. Mean % cover (+ s.e.) of sessile benthos, Bali, April–May 2011 and Nusa Penida (October 2008). HC – Hard Coral; RDC – Recently Dead Coral; ADC – All Dead standing Coral; RBL – coral Rubble; SC – Soft Coral; MA – Macro-Algae; TA – Turf Algae; CA – Coralline Algae.

cover (30 % or more) included Nusa Penida Stations 7.2, 12.1, 12.2 and 13.2, and Bali Stations 4.1, 12.1, 12.2, 13.2 and 28.2 (Appendix 5.2).

Diversity of soft corals and related taxa was moderate to high at some of these sites (Plate 5.5 and see later), although typically dominated by stoloniferous taxa, especially xeniids. There was only low cover of macro-algae at most station, averaging < 2 % overall. Only two sites had moderate MA cover (20 %, Nusa Penida Stations 1.2 and 2.2). Cover of turf and coralline algae was low to moderate overall, averaging 13 % and 9 % cover respectively (Figure 5.5).

5.3.3 Species richness

Bali hosts a rich coral fauna of 406 confirmed hermatypic scleractinian species, of which 367 species were recorded from the main island of Bali and 296 species from Nusa Penida. A further 13 species were recorded during the field surveys but remain unconfirmed at present (Appendix 5.3), such that there are likely to be some 420 hermatypic Scleractinia present, in total. One species, *Euphyllia* spec. nov. is new to science (Plate 5.5), and a second species, *Isopora* sp., may also be undescribed (Plate 5.6), showing significant morphological variation from known species in its genus. Additionally, several widespread species exhibit consistent local morpho-types around Bali.

A further ca. 100 species have distribution ranges that include the general area of the Lesser Sunda Islands (Wallace 1999, Veron 2000, Veron et al. 2009), but were not recorded around Bali or Nusa Penida during the present surveys, possible localized disjunctions related to failures of dispersal and / or recruitment (see later).

Of the 406 confirmed species recorded, almost all are shared with other areas of Indonesia (Appendix 5.3 and see later). The overall high degree of biogeographic similarity notwithstanding, differences exist among these areas in terms of the *relative abundances* of the species present. This in turn has had a differentiating effect on coral community structure (see later).

Plate 5.1. High cover of reef-building corals, Nusa Penida Station N1.2, composed predominantly of *Acropora* spp.

Plate 5.2. High cover of reef-building corals, Bali Station B30.2, composed predominantly by *Porites nigrescens* and *Seriatopora* spp.

Plate 5.3. High cover of reef-building corals, Nusa Penida Station N4.2, composed predominantly by *Acropora* spp. and *Porites* spp.

Plate 5.4. High patchy cover of soft corals, mostly *Sarcophyton* spp., Nusa Penida Station N16.2.

Within-station (point) richness around Bali averaged 112 species (s.d. 42 spp.), ranging from a low of just two species (at site B22, a muddy non-reefal location) to a high of 181 species at B16 (Jemeluk, Amed). Other species-rich sites included Menjangan N (168 spp., site B26) and Penutukang (164 spp., site B21). These results for site and overall richness are similar to those from Bunaken National Park and Wakatobi, higher than for Komodo and Banda Islands, and lower than Raja Ampat, Teluk Cenderwasih, Fak-Fak/Kaimana and Halmahera (Table 5.4).

Other hard corals, soft corals and other biota

In addition to the hermatypic Scleractinia, numerous other hard and soft corals were recorded, with greater or lesser taxonomic certainty (see Methods and Table 5.5). These included 3 species of the ahermatypic dendrophyllid *Tubastrea*, the 'blue coral' *Heliopora coerulea*, 5 species of hydroid 'fire corals' *Millepora*, the 'organ-pipe coral' *Tubibora musica* and lace corals *Stylaster* and *Distichopora* spp., including the recently described *Distichopora vervoorti* Cairns and Hoeksema, 1998 (Table 5.5). An additional 57 genera of alcyonacean soft corals, plus zoanthids, corallimorpharians, hydroids and related sessile benthos were also recorded. In particular, xeniid and neptheid soft coral genera were well represented with high abundance. Diversity and abundance of sponges was also exceptionally high.

Rarity

The Rarity Index, which rated sites in respect of the occurrence of species otherwise rare in the Bali data set, revealed a broad range of RI scores, with site B7 (W Gili Mimpang, Batu Tiga) being most unusual faunistically, followed closely by site B16 (Jemeluk, Amed) (Table 5.6).

Reefs of Menjangan, Penutukang, Sumba Kima and Cenigan channel also scored highly, indicating locally unusual coral composition and abundance (Table 5.6). More than one-quarter of coral species in the total species pool were locally uncommon or rare, occurring in four or less of the 48 sites. Thirty-three reef coral species were recorded from only one site, 41 species from two sites, 22 species from three sites and 26 species from four sites.

5.3.4. Coral Replenishment

Stations with high coral diversity, abundance and live cover were considered important for the maintenance and replenishment of populations. These were ranked using a simple

Table 5.4. Comparison of diversity and other ecological characteristics of Bali with other Indo-West Pacific coral reef areas. KO – Komodo National Park; DE – Derewan, East Kalimantan; W – Wakatobi area, S. Sulawesi; BN – Bunaken National Park; S-T – Sangihe-Talaud Isl.; BRU – Brunei Darussalam; RA – Rajah Ampat area, Papua; BI – Banda Isl., Banda Sea, Maluku. Data from Turak 2002, Turak 2004, Turak 2005, Turak 2006, Turak and DeVantier 2003, Turak and DeVantier 2009 Turak and DeVantier in press,, Turak and Shouhoko 2003, Turak et al. 2003.

Attribute	Bali	KO	DE	W	BN	ST	BI	RA	TC	FFK
Total number of species	406	342	449	396	392	445	301	487	469	469
Average no. of species per site	112	100	164	124	155	100	106	131	178	171
% of sites with over ⅓rd species	38	43	78	41	85	8	61	18	79	65
Average % hard coral cover	28	32	36	32	41	21	40,3	33	27	26
Number of sites surveyed	48	21	36	27	20	52	18	51	33	34
Area covered (× 1000 km^2) approx.	3,7	2	20	10	0,9	23	0,4	30	27	12

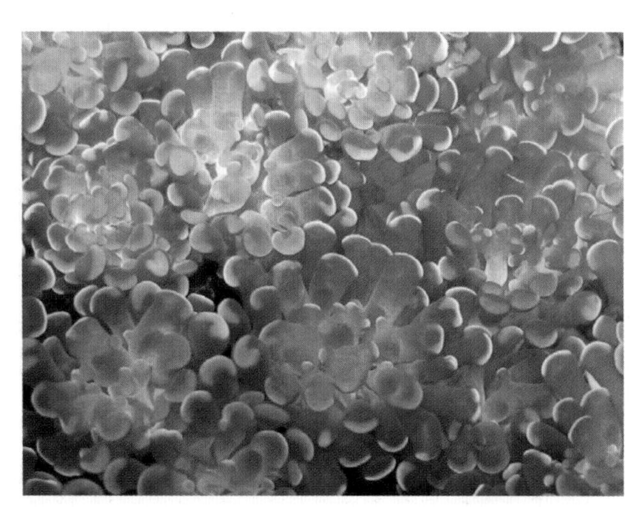

Plate 5.5. Euphyllia spec. nov., discovered by M. Erdmann, E coast of Bali. Close-up of polyp details.

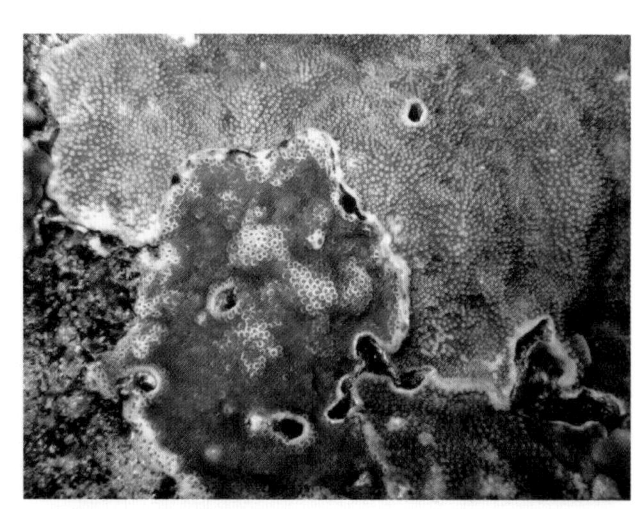

Plate 5.6. Unidentified *Isopora* sp. (center) next to *Isopora palifera* (top and right), Nusa Penida Station N9.2.

Table 5.5. Azooxanthellate scleractinian hard corals, non-scleractinian hard corals, soft corals and other biota recorded in Bali. Results are the number of stations from which each taxon was recorded.

Hard coral taxa	Sites
Scleractinia	
Dendrophylliidae	
Tubastrea micrantha	18
Tubastrea coccinae	14
Tubastrea folkneri	6
Milleporina	
Milleporidae	
Millepora dichotoma	16
Millepora exesa	43
Millepora intricata	10
Millepora platyphylla	18
Millepora tenera	8
Hydroida	
Stylastridae	
Distichopora	4
Stylaster	2
Helioporacea	
Helioporidae	
Heliopora coerolea	18
Alcyonacea	
Tubiporidae	
Tubipora musica	38

Soft coral taxa	Sites
Alcyonacea	
Clavulariidae	
Carijoa	1
Cervera	1
Clavularia	11
Alcyoniidae	
Cladiella	15
Dampia	5
Klyxum	2
Lobophytum	30
Rhytisma	4
Sarcophyton	63
Sinularia spp.	67
Sinularia brascica	
Sinularia flexibilis	10
Nephtheidae	
Capnella	28
Dendronephthya	28
Lemnalia	16
Litophyton	1
Nephthea	40
Paralemnalia	29
Scleronephthya	25
Stereonepthya	2
Umbellulifera	3

Soft coral (*continued*)	Sites
Nidaliidae	
Chironephthya	7
Nephthyigorgia	3
Siphonogorgia	4
Xeniidae	
Anthelia	20
Cespitularia	11
Efflatounaria	13
Heteroxenia	6
Sympodium	3
Xenia	43
Briareidae	
Briareum	28
Anthothelidae	
Alertigorgia	1
Annella	7
Melithaeidae	
Acabaria	1
Melithaea	18
Acanthogorgiidae	
Acanthogorgia	3
Muricella	4
Plexauridae	
Echinogorgia	1
Menella	3
Paraplexaura	1
Villogorgia	1
Gorgoniidae	
Hicksonella	
Pinnigorgia	9
Rumphella	8
Ellisellidae	
Dichotella	5
Elisella	13
Junceella	12
Ifalukellidae	
Ifalukella	1
Isididae	
Isis	3
Pennatulacea	
Veretillidae	
Veretillum	1
Virgulariidae	
Virgularia	2
Pteroeididae	
Pteroeides	1

Other	Sites
Antipatharia	
Antipathidae	
Antipathes	15
Cirrhipathes	19
Zoanthidae	
Palythoa	69
Zoanthus	9
Coralimorpharian	31
Anemon	26
Cerianthus	1
Plumulariidae	2
Aglophenia	30
Lytocarpus philippinus	9

OTHERS	Sites
Spons	31
Cliona	6
Carterospongia	8
Xestospongia	33
Sponge encrusting	25
Sponge massive	17
Sponge blue thin rope	1
Sponge blue tubes	8
Sponge rope	4
Ascidian	
Botryllus	1
Lissoclinum	2
Diademnum	18
Polycarpa	8
Tridacna	
Tridacna crocea	4
Tridacna squamosa	12
Tridacna maxima	17
Echinodermata	
Linckia	18
Culcita	13
Alga	
Halimeda	9
Caulerpa serrulata	7
Dictyosphaeria	15
Turbinaria ornata	12
CRA	33
Peyssonnelia	18
Lamun	
Thalassodendron	3
Halophila ovalis	2
Enhalus	1
Syringodium	1

coral replenishment index CI (Table 5.7 and see Methods). These were widespread across Bali and Nusa Penida, with highest scoring sites at Jemeluk, Amed (B16), Crystal Bay South (N7), Menjangan North (B26) and Toya Pakeh (N3).

Around Nusa Penida in particular, asexual reproduction by fragmentation, budding and/or stoloniferous growth (the latter in soft corals) was particularly prevalent, and may be compensating to some degree for possible low rates of recruitment by planulae, potentially the result of strong current flow limiting local settlement.

Table 5.6. Site ranking (scores) for RI from highest to lowest for top 20 sites, Bali. B indicates site on main island of Bali, N indicates site on Nusa Penida and adjacent smaller islands.

Site name	Site No.	RI
West Gili Mimpang (Batu Tiga)	B7	16,22419
Jemeluk, Amed	B16	14,30168
Menjangan North	B26	11,07563
Penutukang	B21	10,40893
Menjangan East	B28	10,13587
Sumber Kima	B25	10,03883
Ceningen channel	N14	9,164788
Taka Pemutaran	B24	8,910188
Batu Kelibit, Tulamben	B18	8,842868
Kepa, Amed	B17	8,476171
Gili Biaha/Tanjung Pasir Putih	B11	8,115602
Tukad Abu, Tulamben	B19	7,867889
East Gili Mimpang (Batu Tiga)	B9	7,466243
Secret Bay, Reef north shore	B30	7,202368
Gretek	B20	6,80481
Malibu Point	N10	6,659188
Crystal Bay South	N17	6,472372
Gili Selang North	B13	6,375295
Batu Abah	N8	6,288729
South of Batu Abah	N9	6,284583

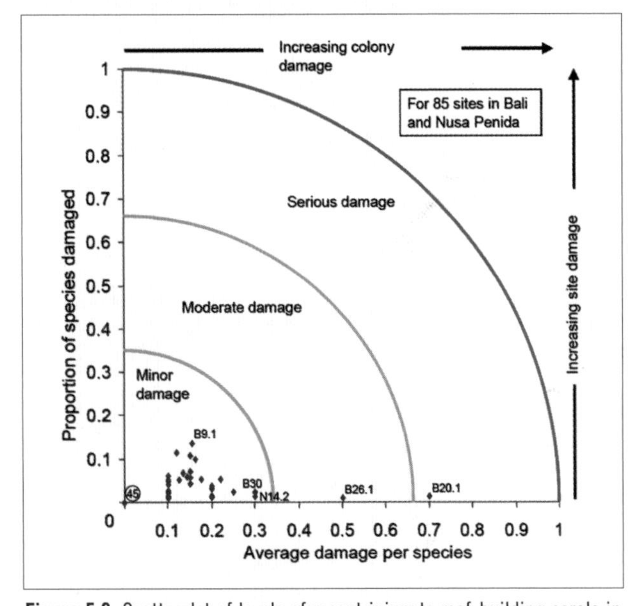

Figure 5.6. Scatterplot of levels of recent injury to reef-building corals in 85 stations, Bali.

5.3.5. Coral injury

Corals of Bali exhibited relatively low levels of recent injury overall, in terms of the proportion of species present that were injured and the average levels of injury to those species (Figure 5.6). This is consistent with the high positive ratio of live : dead coral cover. The overall healthy condition of corals was well represented by the large monospecific stands and intact massive corals present, with little evidence remaining of major past disturbances such as coral bleaching-related mortality triggered by elevated or depressed sea temperatures in 1998, past outbreaks of coral predators, major destructive fishing activities, diseases or other impacts. Earlier impacts had, however, occurred from clearing of corals from reef

Table 5.7. The top 20 sites for the Coral Replenishment index *CI* in Bali. B indicates site on main island of Bali, N indicates site on Nusa Penida and adjacent smaller islands.

Nama Stasiun	No. Stasiun	CI
Jemeluk, Amed	B16	8,46
Crystal Bay South	N7	8,2
Menjangan North	B26	7,95
Toya Pakeh	N3	7,64
Gili Tepekong, Candi Dasa	B10	6,84
Sekolah Dasar	N17	6,63
Mangrove N Lembongan	N4	6,36
Gili Selang South	B14	6,27
Batunggul	N11	6,12
Batu Abah	N8	5,88
Penutukang	B21	5,72
Teluk Lembongan Pantoon	N1	5,72
Bunutan, Amed	B15	5,67
East Gili Mimpang (Batu Tiga)	B8	5,55
Sumber Kima	B25	4,98
Batu Kelibit, Tulamben	B18	4,92
West Gili Mimpang (Batu Tiga)	B7	4,82
Gretek	B20	4,82
Menjangan East	B28	4,7
Gili Biaha/Tanjung Pasir Putih	B11	4,62

Plate 5.7. Seaweed farm, Station N14.2, Nusa Penida

Plate 5.8. Predation by Drupella snails on Acropora yongei, Station N14.1, Nusa Penida.

Plate 5.9. Recent predation by Crown-of-thorns seastars on *Acropora sukarnoi*, Station N8.2, Nusa Penida.

Plate 5.10. Diseased colony of *Goniopora tenuidens*, Station N13.2, Nusa Penida.

Plate 5.11. Blast fishing damage, Station N8.1, Nusa Penida.

flats for development of seaweed farms. The main sources of the relatively minor recent coral injury were predation by *Drupella* snails, and to a lesser extent Crown-of-thorns seastars, and coral diseases (Plates 7–11).

5.3.6 Litter and pollution

Continuing impacts, particularly from litter and other forms of pollution and from poorly regulated / managed tourism development, are of concern. As noted by van Woesik some 15 years ago:

> *"... between September 1992 and September 1997, there has been a major change to the coral reefs at Sanur and Nusa Dua, south-eastern Bali, Indonesia. The reef has changed from being dominated by coral to one being dominated by macroalgae, sponges and filter feeders. This is a sure sign of eutrophication and reef degradation. The source(s) of the eutrophication is presently unknown, but are in urgent need of investigation. Eutrophication is likely to stem from local sewage discharges from Benoa Harbour and local hotels. ... It appears that a priority for south-eastern Bali is to improve the water quality, by investing in sewage treatment"*

The situation in respect of coral – algal cover in 2011 in the Sanur – Nusa Dua area does not appear to have deteriorated further since 1997, although various forms of pollution, most notably plastic and other kinds of litter, were apparent at all sites around the main island of Bali (Plates 5.12–5.13). Sources of the pollution include dumping on the coast and in streams, from ships and boats, and from more distant sources, transported in ocean currents.

One of us (LD) has spent considerable time in Bali since 1975, and personal observations and other anecdotal evidence suggests that the amount of litter, and indeed pollution more generally, has increased significantly over the past several decades, in proportion with Bali's growing human population and the increasingly ubiquitous use of plastics in packaging; and consistent with van Woesik's observations in 1997.

During this study, we had no opportunity to specifically investigate waste treatment and disposal as they relate to water quality on coastal reefs. However, personal observations indicated that in 1975, the only stream that showed obvious signs of pollution was one in the centre of Denpasar. Today, unfortunately, most streams that were crossed during travel to the survey locations around Bali appear polluted to greater or lesser degree by plastics and other forms of waste, much of which is subsequently transported into the coastal marine environment by stream flow. Opportunities exist to reduce these impacts through education programs that encourage continued and expanded use of traditional biodegradable packaging (eg. banana and palm leaf bags), better waste management and disposal, and restoration of riparian zones along streams.

5.3.7 Coral community Structure around Bali

The cluster analysis revealed four main coral community groups (Figure 5.7) at site level, one of which was subdivided into two communities (B and C) based on major differences in exposure, substrate type and other environmental variables (Figures 5.7, 5.8). Each community was characterized by a distinctive suite of species and benthic attributes (Tables 5.8, 5.9, Figure 5.9), although some species were more or less ubiquitous across several community types, notably *Acropora* and *Porites* spp. and various faviids. Because of their commonness, these taxa were not useful in differentiating among the different communities, although they do contribute significantly to coral cover in the region (Plates 5.14–5.23).

Community A: Agariciid – faviid community

This community occurred predominantly in warm waters (mean temperature of 29.6 C) of good clarity (mean visibility of 15 m) along the sheltered N coast of Bali (mean exposure of 2.1), on moderately to well-developed reefs (mean

Plate 5.12. Plastic litter and silt fouling the reef, Bali Station 31.

Plate 5.13. Abandoned 'ghost' net tangling corals, Bali Station B13.2.

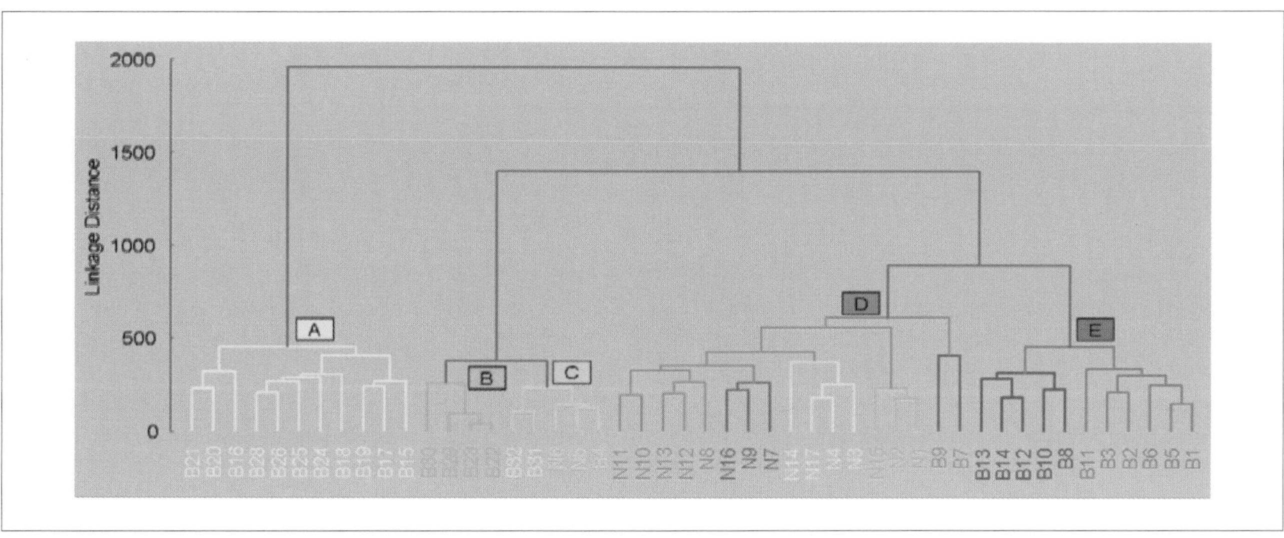

Figure 5.7. Dendrogram illustrating results of cluster analysis of coral communities of 48 sites from Bali (B#) and Nusa Penida (N#.).

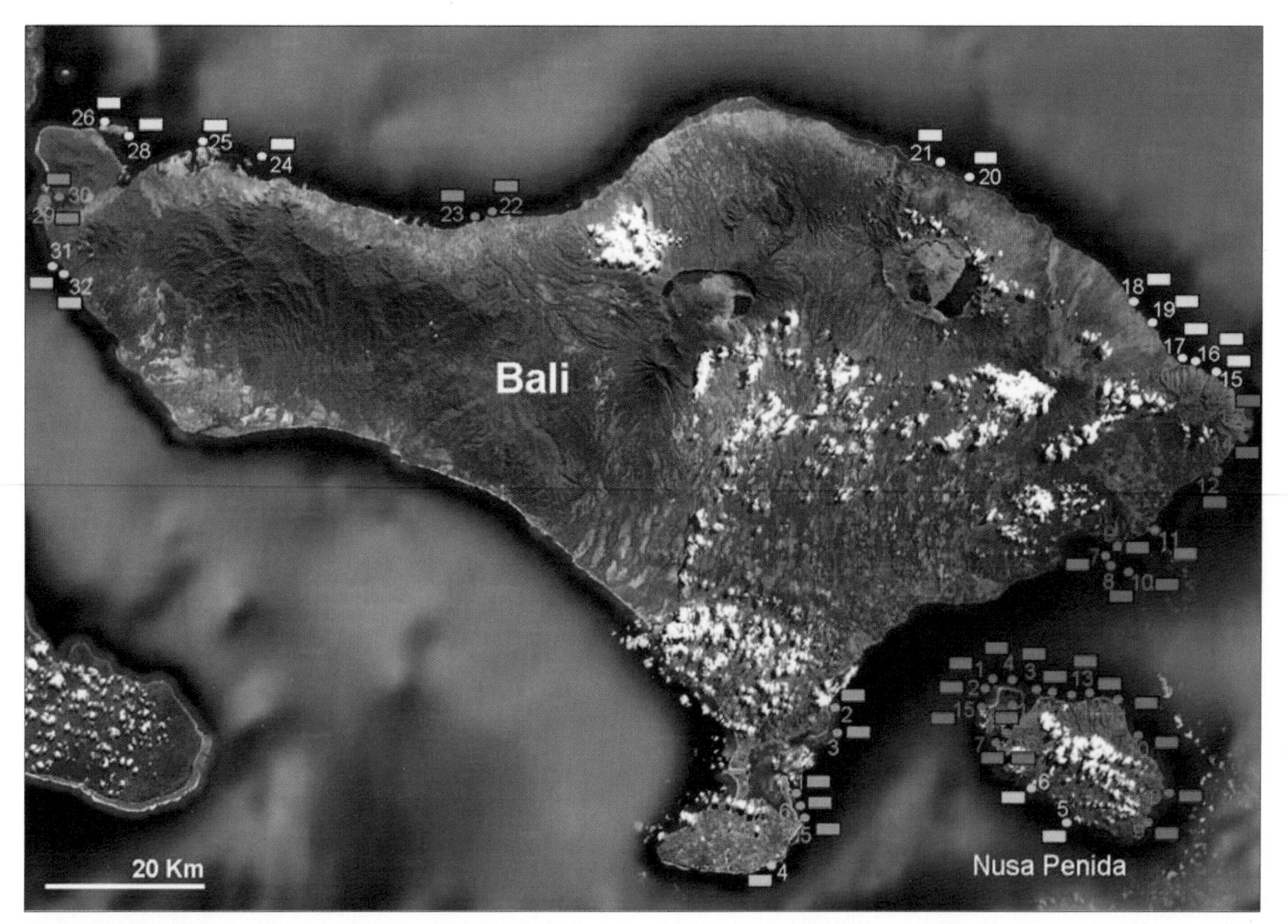

Figure 5.8. Distribution of coral community types at 48 sites, Bali. The 5 communities show a moderately high degree of geographic separation across the survey area. Each site has one shaded 'community rectangle' indicating the identity of the community present, where Community A is represented by yellow, B by brown, C by blue, D by red, and E by pink and purple coloured rectangles.

of 2.5) with relatively steep slopes (mean of 24 degrees) (Figure 5.8, Table 5.8). Characteristic indicator species included the agariciids *Leptoseris explanulata* and *L. mycetoseroides* and *Pavona varians*, faviids *Favites abdita, Favia pallida, Goniastrea retiformis* and *G. aspera* (Table 9). Tabular and branching *Acropora* and foliose *Montipora* were also common. Community A had moderately high cover of living hard corals (mean 28 %) and was the most species rich (mean of 154 reef coral spp. per site) (Plates 5.14, 5.15).

Community B: Pocilloporid – poritid community

This community of Bali's N coast clustered with the following community C in the dendrogram (Figure 5.7) because both communities share low species richness and the presence of stress-tolerant coral species. It was, nevertheless, separated from it based on markedly different environmental characteristics, notably its relatively sheltered exposure regime (mean of 1.8), low degree of reef development (mean of 1.8), low water clarity (mean of 5 m) and very low level of hard substrate (mean of 19 %) and high levels of sand and silt (means of 54 and 25 % repsectively) (Figure 5.6, Table 5.8). This community had moderate cover of live hard corals (22 %) and was characterized by the pocilloporids *Seriatopora, Pocillopora* and *Stylophora* spp., branching and massive poritids *Porites* spp. and the staghorn acroporid *Acropora pulchra* (Table 5.9). The pocilloporids are typically common colonizing species, often with rapid turnover of populations, while massive *Porites* are among the more stress-tolerant. The small free-living corals *Heterocyathus* and *Heteropsammia*, and seagrass *Halophila ovalis*, all typically associated with soft sediments, were also present. Community B had lowest species richness (mean of 19 reef coral spp. per site) (Plates 5.16, 5.17). These various environmental and biotic characteristics are consistent with a community of marginal and/or stressed reef habitats.

Community C: Faviid – pectiniid community

This community occurred in the more exposed locations of the S coasts of Bali and Nusa Penida, extending along the W coast towards the NW corner of Bali (mean exposure of 2.8), in cooler waters (mean temperature of 27.8 C) of low clarity (mean of 7 m), and with low levels of reef development (mean of 1.8) (Table 5.8, Figure 5.8). It was characterized by a mix of massive and encrusting faviids *Favia, Favites, Platygyra, Plesiastrea, Cyphastrea* and *Echinopora* and the encrusting-plating pectiniids *Mycedium elephantotus* and *Oxypora lacera* (Table 5.9). Community C had low reef coral species richness (mean of 56 spp. per site), lowest mean cover of living hard corals (mean 12 %) and highest cover of soft corals (mean of 18 %) (Plates 5.18, 5.19). It is likely that this community is widespread along the more wave-exposed coastlines of Bali and Nusa Penida.

Community D: Mussid – merulinid community

This community occurred predominantly along the more wave-sheltered coastlines of Nusa Penida and adjacent islands, and to a lesser extent on the E coast of Bali (Figure 5.8), on well developed reefs (mean of 3.2) in areas of high water clarity (mean of 19 m) and often in areas of moderate to strong current flow. It had highest cover of living hard corals (mean of 36 %) and moderately high cover of soft corals (mean of 11 %), and was moderately species rich (mean of 117 reef coral species per site, Table 5.8), being characterized by the mussids *Lobophyllia* and *Symphyllia* spp. and merulinids *Hydnophora* and *Merulina* spp. (Table 5.9, Plates 5.20, 5.21). Tabular and branching *Acropora* and foliose *Montipora* were also common. This community had highest cover of rubble and dead coral (means of 11 % and 3 % respectively). As illustrated in the dendrogram (Figure 5.7), this community encompasses four of the five coral communities previously identified for the Nusa Penida area in the earlier, smaller stand-alone analysis focused on that

Table 5.8. Summary statistics (mean values) for environmental and benthic cover variables for 5 coral communities of Bali. Differentiating characteristics are in bold type.

Coral community attributes					
	A	B	C	D	E
No. of sites	11	4	4	17	11
Maximum depth (m)	20	16	20	19	14
Minimum depth (m)	6	1	7	5	5
Slope (degrees angle)	24	8	9	13	17
Hard substrate (%)	73	**19**	**87**	80	82
% cover benthos					
Hard coral	28	22	**12**	**35**	26
Soft coral	5	0	**18**	11	12
Macro algae	1	1	4	2	2
Turf algae	17	**3**	17	10	13
Coralline algae	10	0	2	8	**13**
Recently dead coral	1	0	0	1	1
All dead coral	2	0	2	3	2
% cover substrate					
Continuous pavement	50	15	75	62	54
Large blocks	12	0	8	11	16
Small blocks	11	4	6	7	11
Rubble	9	3	2	11	5
Sand	15	**54**	10	9	14
Silt	3	**25**	0	0	0
Environmental variables					
Exposure	2.1	**1.8**	**2.8**	2.4	2.6
Reef development	2.5	1.8	1.8	**3.2**	2.8
Visibility (m)	15	5	7	**19**	13
Water temp (C)	29.6	28.5	27.8	28.1	28.6
Mean no. of reef-building coral species	**154**	**19**	59	117	119

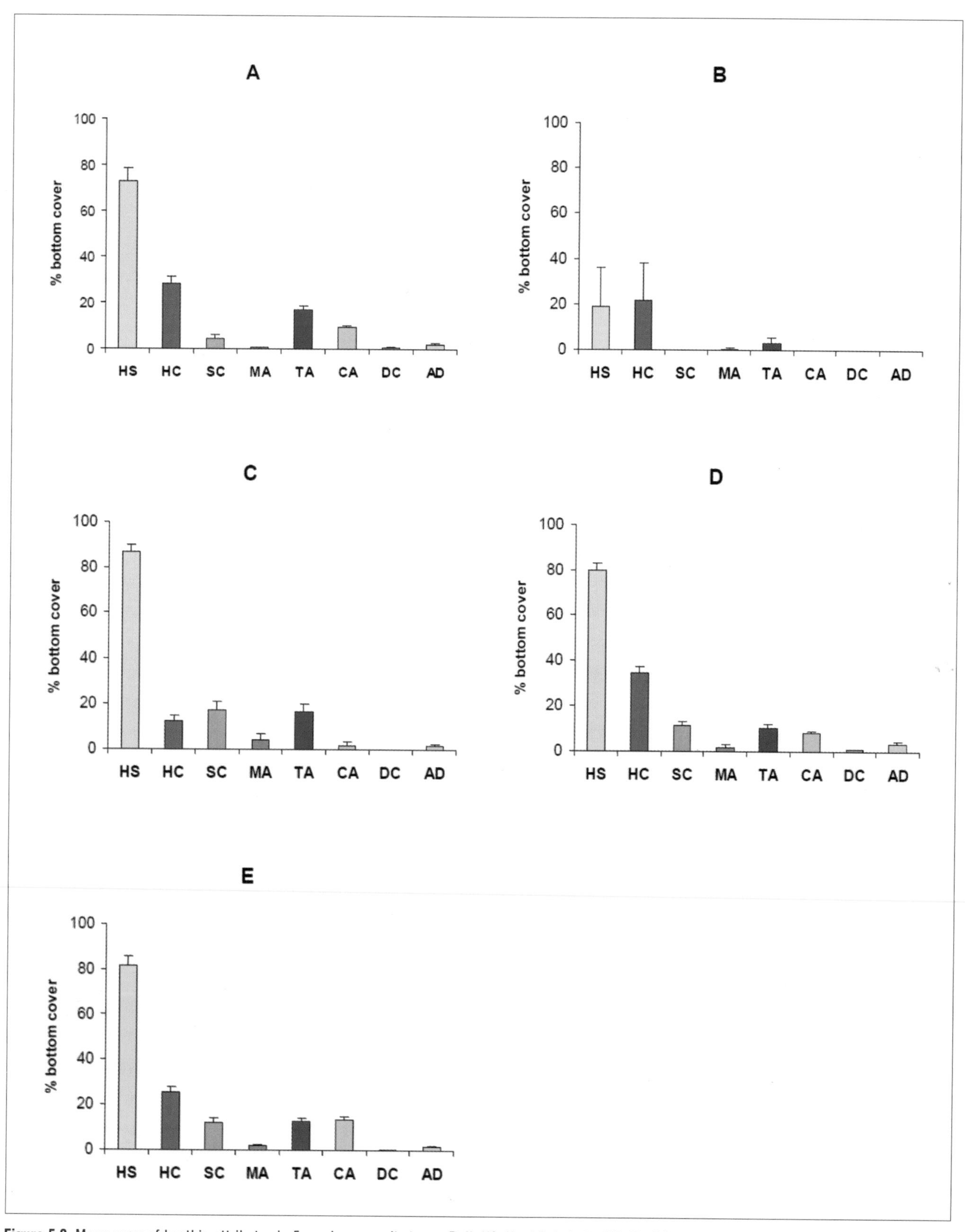

Figure 5.9. Mean cover of benthic attributes in 5 coral community types, Bali. HS: Hard Substrate, HC: Hard Corals, SC: Soft Coral, MA: Macro Algae, TA: Turf Algae, CA: Coralline Algae, DC: Recently Dead Coral: AD: Old Dead Coral. Error bars are Standart Error (SE).

Table 5.9. Characteristic coral species in 5 coral community types, Bali. Taxa used as indicators for the relevant community types are in bold.

Community A			**Community B**		
Scleractinia	abn	stn	Scleractinia	abn	stn
Leptoseris explanata	27	11	*Porites* massive	5	3
Porites massive	26	11	*Seriatopora hystrix*	6	2
Pocillopora verrucosa	24	11	*Porites nigrescens*	6	2
Favites abdita	24	11	*Seriatopora caliendrum*	4	2
Porites cylindrica	24	11	*Stylophora pistillata*	4	2
Montipora grisea	23	11	***Acropora pulchra***	4	2
Pavona varians	23	11	*Hydnophora rigida*	4	2
Galaxea fascicularis	22	11	***Pavona decussata***	3	2
Favia pallida	22	11	*Cyphastrea serailia*	3	2
Goniastrea retiformis	22	11	***Heterocyathus aequicostatus***	4	1
Platygyra daedalea	22	11	*Pocillopora damicornis*	3	1
Favites pentagona	21	11	*Euphyllia paraancora*	3	1
Goniastrea pectinata	21	11	*Heteropsammia cochlea*	3	1
Leptoseris mycetoseroides	20	11	*Porites flavus*	3	1
Goniastrea aspera	20	11	*Goniopora stokesi*	3	1
Montastrea colemani	20	11	*Stylophora subseriata*	2	1
Porites rus	20	11	*Montipora aequituberculata*	2	1
Acropora tenuis	19	11	*Montipora altasepta*	2	1
Hydnophora microconos	19	11	*Montipora delicatula*	2	1
Symphyllia recta	19	11	*Acropora tenuis*	2	1
Other taxa			**Other taxa**		
Palythoa	23	11	**Pennatulacea**	4	2
Sinularia spp.	19	11	*Caulerpa taxifolia*	4	2
Sarcophyton	14	10	***Halophila ovalis***	4	2
Sponge massive	21	9	*Culcita*	3	2
Dendronephthya	18	9	Sponge	3	1
Xestospongia	16	8	*Millepora exesa*	2	1
Millepora exesa	15	8	*Millepora intricata*	2	1
Linckia	14	8	*Clavularia*	2	1
Sponge encrusting	14	7	*Dendronephthya*	2	1
CRA	14	7	*Xenia*	2	1
Carterospongia	13	7	*Antipathes*	2	1
Melithaea	12	7	Sponge rope	2	1
Lobophytum	11	7	*Padina*	2	1
Tridacna maxima	10	7	*Caulerpa serrulata*	2	1
Sponge	15	6	*Caulerpa racemosa*	2	1
Diademnum	14	6	*Syringodium*	2	1
Polycarpa	11	6	*Lobophytum*	1	1
Tridacna squamosa	7	6	*Heteroxenia*	1	1
Culcita	7	6	*Zoanthus*	1	1
Aglophenia	12	5	Anemon	1	1

table continued on next page

Table 5.9. *continued*

Community C		
Scleractinia	abn	stn
Favites pentagona	12	5
Galaxea fascicularis	11	5
Platygyra daedalea	10	5
Plesiastrea versipora	10	5
Symphyllia recta	8	5
Favia speciosa	8	5
Pachyseris speciosa	7	5
Mycedium elephantotus	7	5
Cyphastrea serailia	6	5
Acropora sukarnoi	8	4
Oxypora lacera	8	4
Hydnophora exesa	8	4
Favites russelli	8	4
Leptoseris explanata	7	4
Pocillopora eydouxi	6	4
Echinopora lamellosa	6	4
Symphyllia agaricia	5	4
Symphyllia valenciennesii	5	4
Favia favus	5	4
Porites massive	5	4
Other taxa		
Sinularia spp.	13	5
Sarcophyton	9	4
Xestospongia	8	4
Capnella	6	4
Junceella	6	4
Palythoa	6	4
Lobophytum	7	3
Sponge	7	3
Tubipora musica	6	3
Dampia	6	3
Xenia	6	3
Anemon	4	3
Tubastrea micrantha	6	2
Amphiroa	6	2
Aglophenia	5	2
Coralimorpharian	4	2
Dictyosphaeria	4	2
Nephthea	3	2
Melithaea	3	2
Elisella	3	2

Community D		
Scleractinia	abn	stn
Galaxea fascicularis	40	17
Favites pentagona	33	17
Platygyra daedalea	33	17
Pavona explanulata	28	17
Lobophyllia hemprichii	27	17
Symphyllia recta	26	17
Goniopora tenuidens	43	16
Echinopora lamellosa	37	16
Porites massive	32	16
Pavona varians	31	16
Hydnophora exesa	30	16
Acropora microclados	25	16
Symphyllia agaricia	21	16
Lobophyllia robusta	20	16
Pocillopora verrucosa	31	15
Pectinia lactuca	27	15
Merulina scabricula	25	15
Favia favus	23	15
Favia matthaii	22	15
Symphyllia radians	19	15
Other taxa		
Sarcophyton	34	17
Xenia	45	16
Sinularia spp.	31	16
Palythoa	31	16
Millepora exesa	30	16
Tubipora musica	32	15
Coralimorpharian	30	15
Capnella	26	13
Nephthea	26	13
CRA	27	12
Paralemnalia	25	12
Anthelia	25	12
Anemon	19	12
Sponge	27	11
Briareum	19	11
Cirrhipathes	15	11
Scleronephthya	23	10
Lemnalia	19	10
Xestospongia	18	10
Dictyosphaeria	17	9

table continued on next page

Table 5.9. *continued*

Community E		
Scleractinia	**abn**	**stn**
Favites pentagona	29	11
Seriatopora hystrix	27	11
Porites cylindrica	27	11
Pocillopora verrucosa	23	11
Acropora sukarnoi	23	11
Favia matthaii	22	11
Echinophyllia aspera	21	11
Favia favus	21	11
Lobophyllia hemprichii	20	11
Favia speciosa	20	11
Platygyra daedalea	20	11
Pocillopora eydouxi	19	11
Acropora microclados	19	11
Symphyllia agaricia	19	11
Plesiastrea versipora	19	11
Symphyllia radians	17	11
Symphyllia recta	17	11
Porites massive	17	11
Favites russelli	15	11
Montipora vietnamensis	13	11
Other taxa		
Sarcophyton	25	11
Sinularia spp.	25	11
Aglophenia	22	10
Palythoa	21	10
Xenia	23	9
Lobophytum	16	9
CRA	20	8
Xestospongia	15	8
Tubipora musica	15	7
Nephthea	14	7
Peyssonnelia	14	6
Capnella	13	6
Briareum	10	6
Millepora exesa	9	6
Paralemnalia	11	5
Sponge encrusting	11	5
Cespitularia	9	5
Millepora platyphylla	7	5
Sponge massive	7	4
Millepora dichotoma	6	4

area (Turak and DeVantier 2008), prior to addition of the Bali dataset in this larger analysis.

Community E: Acropora sukarnoi community
This community occurred exclusively along the E coast of Bali (Figure 5.8). It can be separated into two sub-communities (illustrated in Figure 5.8 with pink and purple rectangles respectively), the former mainly occurring on the SE coast in the Nusa Dua – Sanur area, the latter further to the NE, around Candi Dasi – Padang Bai – Talumben. This community had moderately high cover of living hard and soft corals (means of 26% and 12% respectively) and was moderately species rich (mean of 119 reef coral species per site, Table 5.8), being characterized by the acroporids *Acropora sukarnoi* and *A. microclados* and *Montipora vietnamensis*, poritid *Porites cylindrica* and pocilloporid *Pocillopora eydouxi* (Table 5.9, Plates 5.22, 5.23).

5.3.8 Comparisons between Bali and adjacent regions
Bali shares almost all coral species with other areas of Indonesia (Appendix 5.3), with the possible exception of *Acropora suharsonoi* (Plate 5.24) and the undescribed species of *Euphyllia* found during the present study (Plate 5.6). Comparisons of levels of similarity in coral composition and community structure were conducted with those of other regions of Indonesia, including Komodo, Wakatobi, Derewan, Banda Islands, Bunaken, N Halmahera and three areas of the Bird's Head Seascape (Raja Ampat, Teluk Cenderwasih and Fak-Fak/Kaimana)

For these regional comparisons, two sets of analyses were undertaken:

1. Using the presence of species in each region
2. Using the species-abundance at the individual site level:
    a. For Bali with Komodo, Banda Islands, Bunaken and Wakatobi (134 sites)
    b. For Bali with Derewan, Sangihe-Talaud, Raja Ampat, Fak-Fak/Kaimana and Teluk Cenderwasih (254 sites).

1. Species presence
Corals of Bali and Nusa Penida were most similar to those of Komodo, the closest location geographically, also forming part of the Lesser Sunda Island chain, and also subject to localized cool water upwelling. These two locations formed a second cluster with Wakatobi and Bunaken, and then with Banda Islands (Figure 5.10).

A second major group of locations included Derewan, Sangihe-Talaud, Halmahera, Raja Ampat, Fak-Fak/Kaimana and Teluk Cenderwasih, reflecting their higher overall species richness (and habitat diversity).

2. Species – abundance
Most sites from Bali and Nusa Penida formed one or more coherent sub-clusters (Figs. 11 and 12, illustrated in purple and pink). Coral communities of Bali and Nusa Penida were

Plate 5.14. Example of coral community A, Station B16.2, Bali, here showing very high cover of reef corals in shallow waters, mainly acroporids *Montipora* (foreground) and *Acropora*.

Plate 5.15. Example of coral community A, Station B17.1, Bali, showing impact of silt.

Plate 5.16. Example of coral community B, Station B30.2, Bali, composed predominantly of *Acropora pulchra* with smaller *Seriatopora hystrix*.

Plate 5.17. Example of coral community B, Station B22.2, Bali, with many small, unattached *Heterocyathus* and *Heteropsammia* corals scattered among seagrass *Halophila* on a soft substrate.

Plate 5.18. Example of coral community C, Station B5.1, Nusa Penida, with encrusting and plating pectiniids and faviids predominant.

Plate 5.19. Example of coral community C, Station B4.1, Bali, with rhodophyte algae and soft corals predominant.

most similar to each other and then with those of Komodo (and a few Banda Island sites). These collectively formed one of the two main clusters of sites (left of Figure 5.11). The second main community group cluster was composed predominantly by sites from Wakatobi, Banda Islands and Bunaken, with some sites from N Bali sharing similarity with some Bunaken sites.

In the second site-level analysis (Figure 5.12), there is a clear clustering of coral communities of Bali with those of Nusa Penida, forming coherent sub-clusters in the large community grouping (left of Figure 5.12). Other more-or-less coherent sub-clusters were formed by sites of Fak-Fak/Kaimana; Derewan and Raja Ampat (in part) and Sangihe-Talaud (with some RA sites). Various sites from Teluk Cenderawasih were spread widely across the dendrogram, some clustering with Fak-Fak/Kaimana, others with Derawan and Raja Ampat (Figure 5.12).

These various results indicate that in terms of both coral species composition (presence, Figure 5.10) and abundance (community structure, Figures 5.11 and 5.12), Bali and Nusa Penida have a degree of self-similarity with each other and dissimilarity from most other regions of Indonesia, being closest in these attributes to Komodo, also in the Lesser Sunda Island chain.

Plate 5.20. Example of coral community D, Nusa Penida Station N1.2, composed predominantly of tabular and foliose acroporids.

Plate 5.21. Example of coral community D, Nusa Penida Station N8.2, showing a diverse coral assemblage developed on an irregular reef spur.

Plate 5.22. Example of coral community E, Station B6.2, Bali, with large stand of *Acropora sukarnoi* (centre).

Plate 5.23. Example of coral community E, Station B8.2, Bali, with large tabular *Acorpora cytherea* (centre).

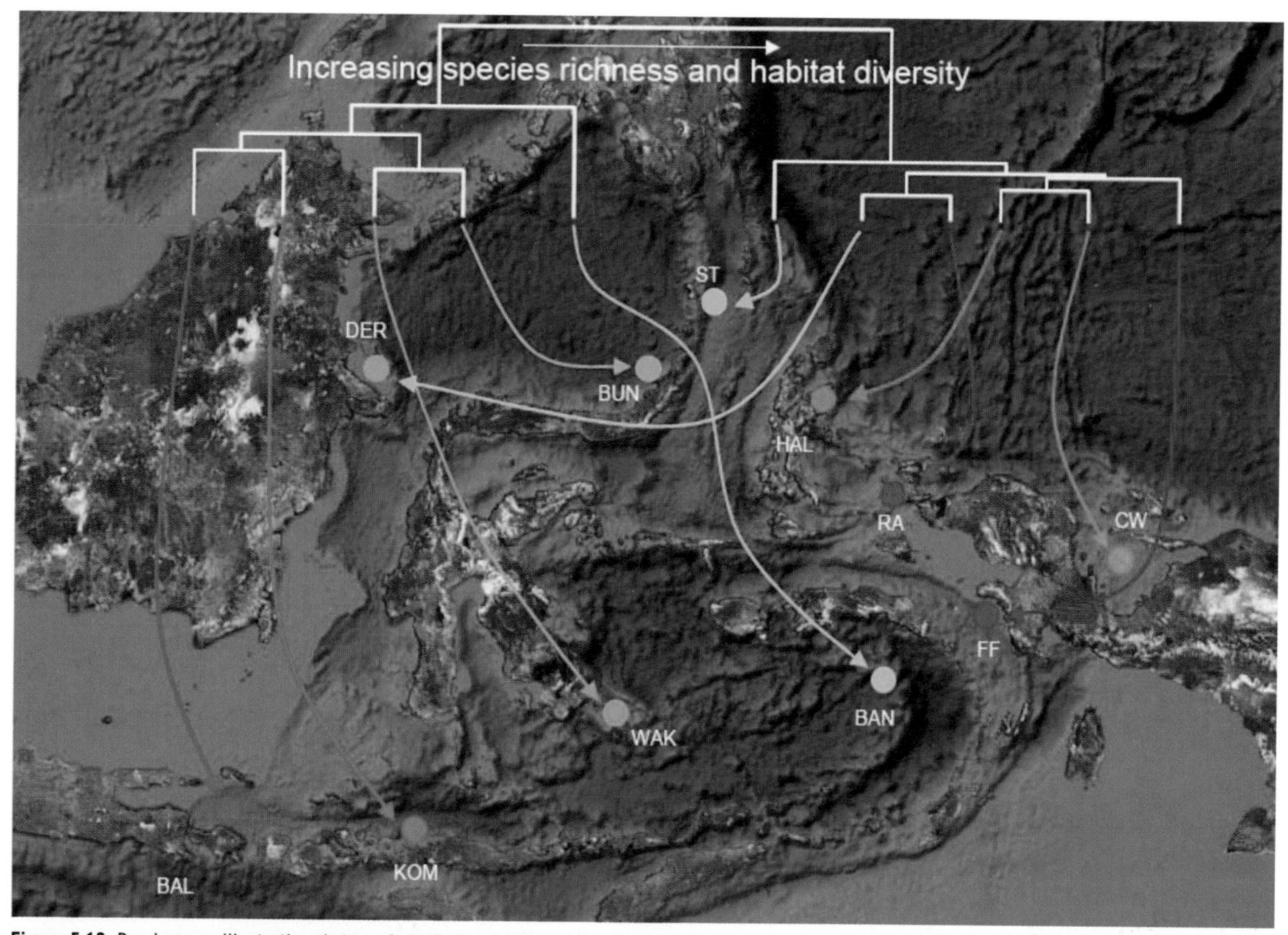

Figure 5.10. Dendrogram illustrating degree of similarity of different locations in terms of reef coral species presence, where BAL – Bali and Nusa Penida, KOM – Komodo, WAK – Wakatobi, BUN – Bunaken, BAN – Banda Islands, DER – Derewan, ST – Sangihe-Talaud, HAL – Halmahera, RA – Raja Ampat, FF – Fak-Fak/Kaimana and CW –Teluk Cenderawasih.

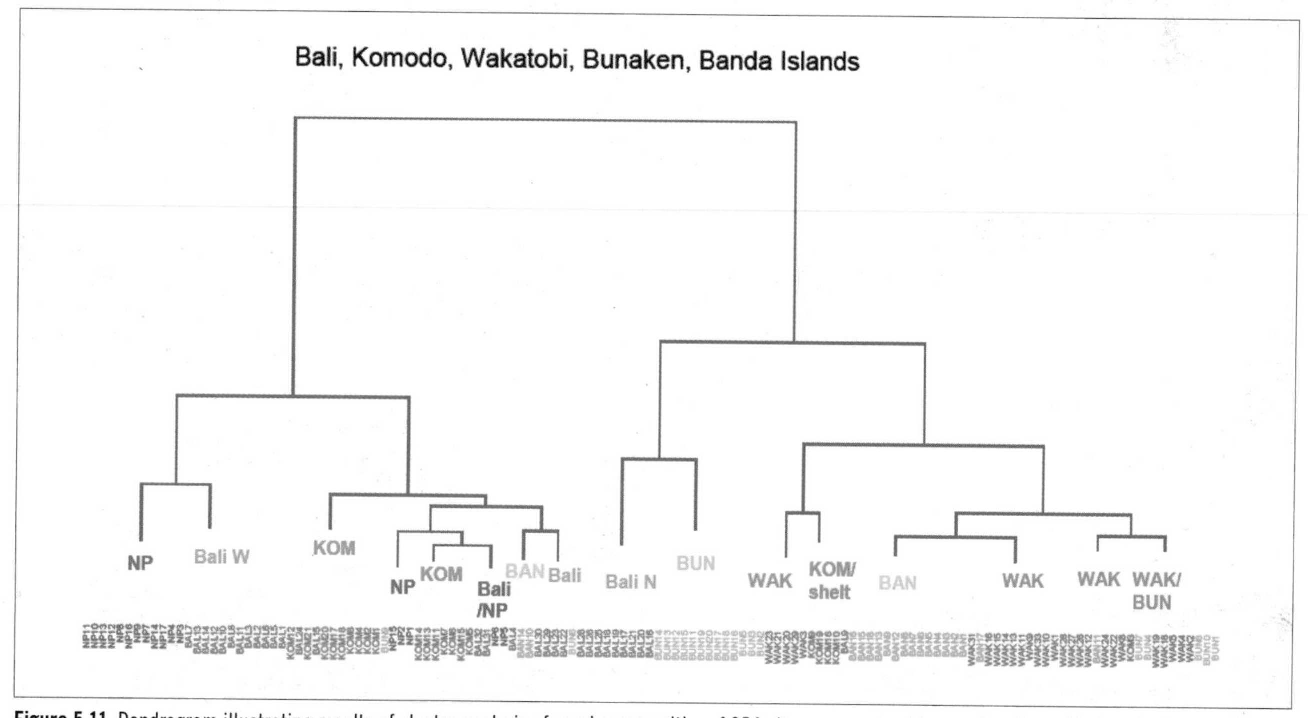

Figure 5.11. Dendrogram illustrating results of cluster analysis of coral communities of 254 sites across six widespread regions of Indonesia: Bali (with geog. location of sites), Nusa Penida (NP), Komodo (KOM), Bunaken (BUN), Wakatobi (WAK) and Banda Islands (BAN).

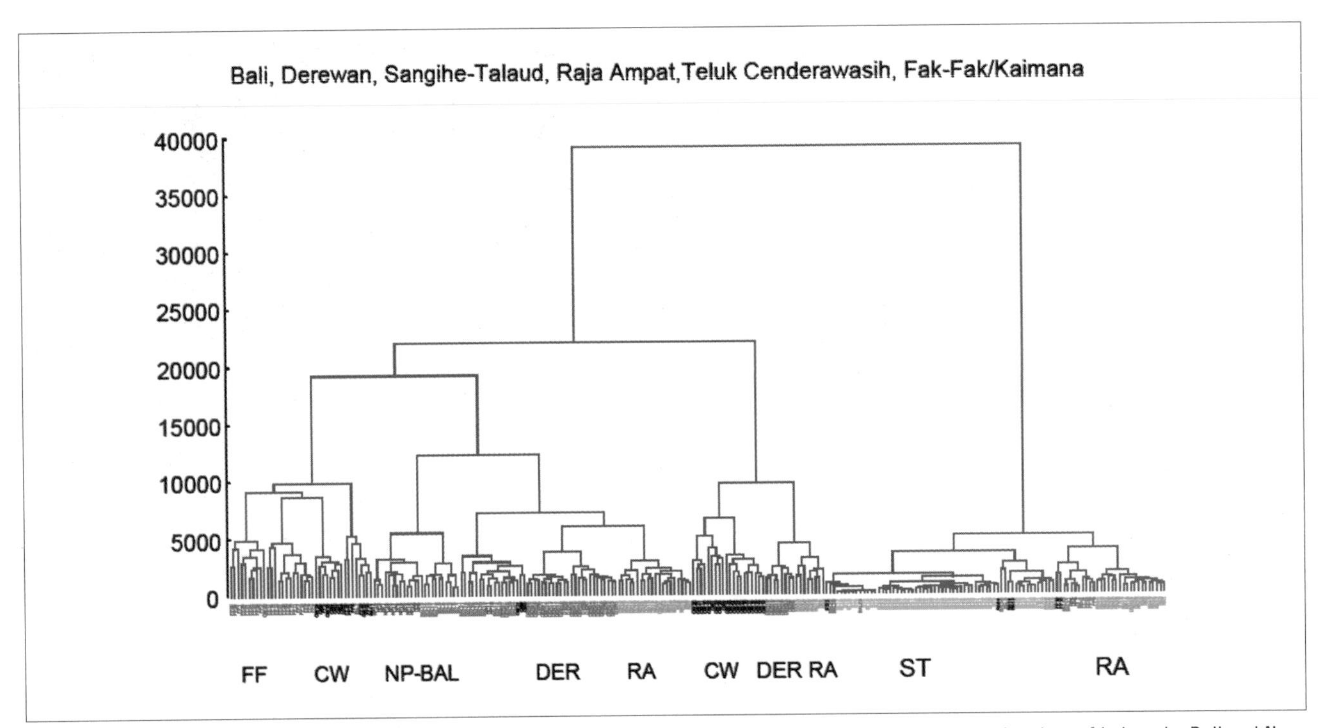

Figure 5.12. Dendrogram illustrating results of cluster analysis of coral communities of 254 sites across six widespread regions of Indonesia: Bali and Nusa Penida (NP-BAL), Derewan (DER), Fak-Fak/Kaimana (FF), Teluk Cenderawasih (CW), Raja Ampat (RA) and Sangihe Talaud (ST).

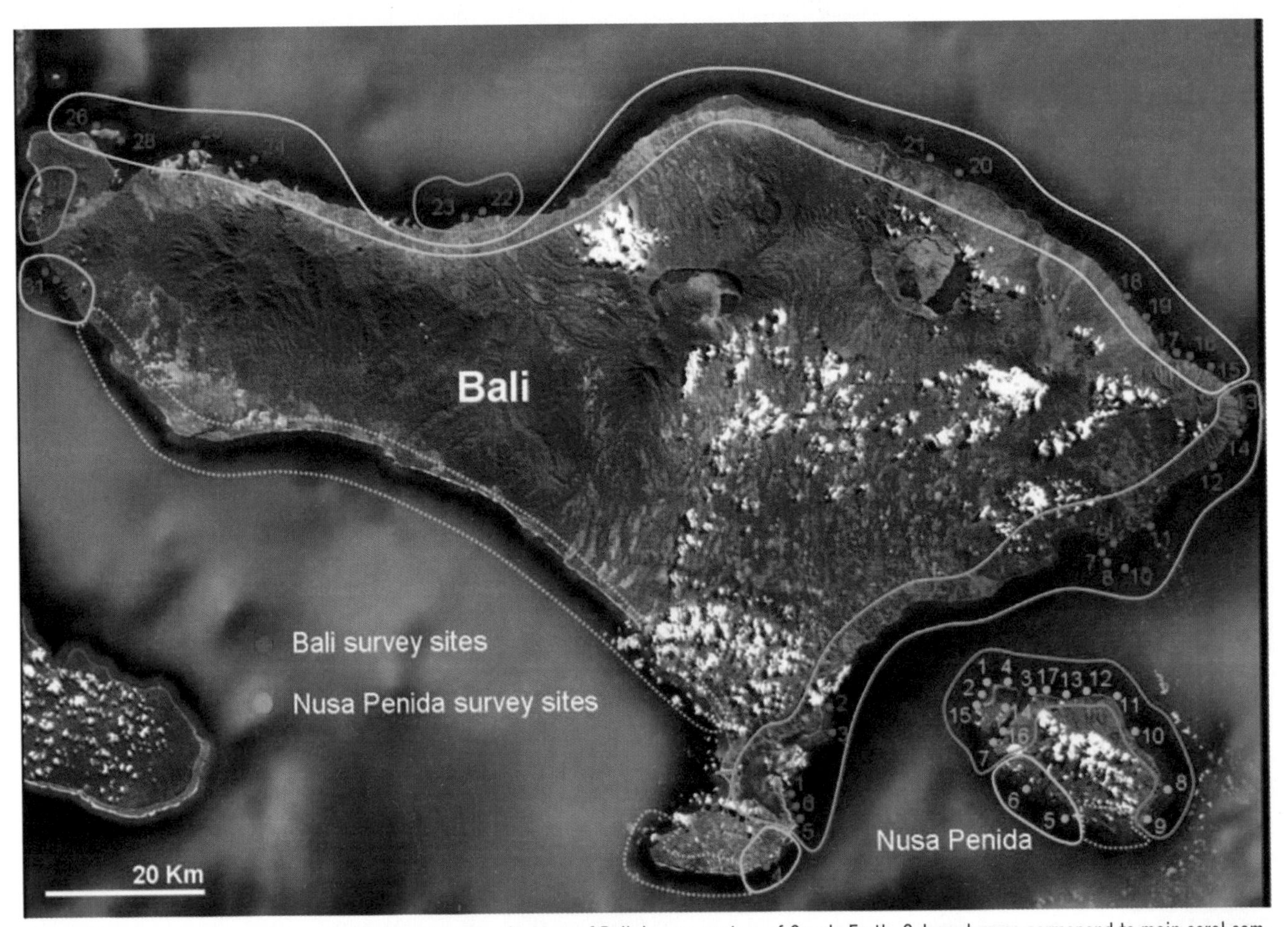

Figure 5.13. Areas hosting main coral habitats and community types of Bali. Image courtesy of Google Earth. Coloured areas correspond to main coral community types in Figure 5.7.

5.4 DISCUSSION

Bali's coral habitats, although not as diverse as some other regions of Indonesia, do encompass a broad range of environmental conditions, and can be differentiated based on the following characteristics (after DeVantier et al. 2008):

1. S coasts – upwelling and / or wave exposed
2. Lombok Strait – variable temperature regime and strong current flow, with some areas likely to be biologically isolated via the strong ITF currents
3. NE and N coasts – warmer waters and more sheltered, mix of soft and hard substrates
4. NW coast – best reef development, but also with significant areas of soft substrate (eg. sites 22, 23, 29, 30).

These habitat types have a major structuring influence on coral communities, as simplified in Figure 5.13. Planning for a network of Marine Protected Areas (MPAs) should aim to represent these main habitat / community types in the network, with particularly important reef stations in each habitat highlighted later herein (Table 5.10, Figure 5.14).

Bali's regional richness of 406 reef coral species is higher than Komodo (342 spp., also in the Lesser Sunda Islands), and Banda Islands (301 spp.) and very similar to Bunaken NP and Wakatobi (Table 5.4). Bali's species composition and community structure show closest similarity with Komodo (Figures 5.10, 5.11), reflecting the fact that both regions experience a similar range of physico-chemical environmental conditions in respect of the sea temperature regime (localized cool water upwelling), current flow and attenuation in wave energy around the islands. Further afield, species composition of Bali differs substantially from the more species rich (and habitat rich) regions of Derewan, Sangihe-Talaud and the Bird's Head Seascape of West Papua (Figures 5.10, 5.12).

Discovery of an undescribed species of *Euphyllia* on the E coast of Bali (Plate 5.5), and the presence of other apparently local endemic corals, notably *Acropora suharsonoi* (Plate 5.24), suggests that the region does have a degree of faunal uniqueness, possibly related to the strong current flow through Lombok Strait. In this respect, the strong ITF currents may, paradoxically, both limit and foster dispersal and recruitment in different areas respectively. Local recruitment around Nusa Penida may be restricted by the currents, which may nevertheless transport larvae further afield. Genetic, reproductive and larval settlement studies would be required to test this hypothesis.

Plate 5.24. *Acropora suharsonoi*, a reef coral with an apparently highly restricted distribution range of N Bali and W Lombok, here at Bali site B26.

5.4.1 Conservation priority

Nusa Penida

The high coral cover of many sites around Nusa Penida in particular may be maintained more by asexual reproduction and growth of fragments, evidence for which was apparent in the large monospecific coral stands and spread of stoloniferous soft corals. Based on known rates of growth, the largest monospecific stands (eg. *Acropora horrida*) are likely to be 100s of years old, and play important roles in maintenance of community structure, providing a high degree of ecological stability to their stations. The presence of local morpho-types in several widespread coral species around Nusa Penida, and apparent absence of species known from adjacent areas (eg. *Acropora surharsonoi* (Plate 5.24) from the Gilis, Lombok and NE Bali mainland) lends further support to the likely role of the ITF in isolating the Nusa Penida islands from other sources of replenishment, both local and further afield. If this is the case, then the islands may require careful management of local impacts, as replenishment from outside sources may be a prolonged process.

Locally, most coral communities of Nusa Penida differ from those of the main island of Bali (Figures 5.7–5.13), and are subject to different environmental conditions and human uses, and hence may require separate management focus. Reefs of high local conservation status around Nusa Penida include those at Crystal Bay, Toya Pakeh, Sekolah Dasar and N Lembongan (Stations N3, N4, N7, N8, N14 and N17, Table 5.10, Figure 5.14). Although all these sites were lumped in major community type D in the broader analysis that included all the Bali sites (Figure 5.7), they do support several different coral assemblage types, designated as different coloured sub-clusters in Figure 5.7 (and as presented in Turak and DeVantier 2009).

Bali

Reefs of high conservation value around Bali were widespread along the E and N coasts, and include Jemeluk, Menjangan, Gili Tepekong, Penutukang, Bunutan, Gili Selang and Gili Mimpang (Stations B16, B26, B10, B14, B21, B15, B25, B8, B18 and B7), representing mainly community types A and E.

Along with the Nusa Penida reefs already identified (Community type D), all the above reefs have strong potential for development of MPAs providing sufficient logistic resources and long-term support are provided. Notably, site 26 at Menjangan already forms part of the marine protected area

Table 5.10. Conservation values of survey sites in Bali. Replenishment Index (CI) scores from highest to lowest; Rarity Index (RI) ranked from highest (1, most unusual faunistically) to lowest. Species richness – reef-building Scleractinia; site numbers and community types correspond with those in Figures.

Site name	Site No.	CI	RI	HC cover	Species richness	Community type
Jemeluk, Amed	B16	8,46	2	32,5	181	A
Crystal Bay South	N7	8,2	25	55	123	D
Menjangan North	B26	7,95	3	50	168	A
Toya Pakeh	N3	7,64	33	55	114	D
Gili Tepekong, Candi Dasa	B10	6,84	23	40	137	E
Sekolah Dasar	N17	6,63	17	45	138	D
Mangrove N Lembongan	N4	6,36	22	45	134	D
Gili Selang South	B14	6,27	29	32,5	125	E
Batunggul	N11	6,12	21	35	140	D
Batu Abah	N8	5,88	19	50	121	D
Penutukang	B21	5,72	4	27,5	164	A
Teluk Lembongan Pantoon	N1	5,72	39	60	81	D
Bunutan, Amed	B15	5,67	28	32,5	120	A
East Gili Mimpang	B8	5,55	35	32,5	122	E
Sumber Kima	B25	4,98	6	30	154	A
Batu Kelibit, Tulamben	B18	4,92	9	30	157	A
West Gili Mimpang	B7	4,82	1	27,5	142	D
Gretek	B20	4,82	15	20	150	A
Menjangan East	B28	4,7	5	20	150	A
Gili Biaha/Tanjung Pasir Putih	B11	4,62	11	17,5	142	E
Tukad Abu, Tulamben	B19	4,56	12	21	156	A
Kepah, Amed	B17	4,54	10	22,5	158	A
Malibu Point	N10	4,38	16	30	141	D
Glady Willis, Nusa Dua	B2	4,32	27	22,5	133	E
Jepun, Amuk Bay, Candi Dasa	B9	4,04	13	12,5	126	D
Ceningen channel	N14	4,04	7	20	119	D
Taka Pemutaran	B24	3,96	8	25	138	A
Gili Selang North	B13	3,88	18	27,5	117	E
Sental	N13	3,86	24	20	126	D
Seraya	B12	3,84	31	30	110	E
Terora, Sanur	B1	3,82	26	22,5	126	E
Mushroom Bay North	N2	3,75	40	50	74	D
Melia Bali hotel	B6	3,74	32	27,5	121	E
South of Batu Abah	N9	3,62	20	17,5	116	D
Buyuk	N12	3,62	34	30	115	D
Secret Bay, Reef north shore	B30	3,36	14	60	44	B
Crystal Bay Rock	N16	3,34	30	25	103	D
Mushroom Bay South	N15	2,64	36	30	81	D
Sanur Channel N side	B3	2,46	37	20	79	E
Pearl farm, NW Bali	B31	2,18	41	20	75	C
Nusa Dua – Public beach	B5	1,51	38	10	102	E
Manta Point	N5	1,04	42	10	70	C
Pura Kutuh	B4	0,8	46	10	62	C
Peternakan mutiara, Barat laut Bali	B32	0,72	47	10	45	C
Old Manta Bay	N6	0,71	44	3	42	C
Secret Bay, Muck dive	B29	0,36	45	3	21	B
Kalang Anyar	B23	0,1	48	1	8	B
Puri Jati	B22	0,07	43	2	2	B

of Bali Barat. Reefs at Jemeluk (Amed) and around Gili Tepekong, Gili Selang and Gili Mimpang are also of very high conservation value for a number of different criteria (Table 5.10). The Batu Tiga area in particular has strong potential for development of an MPA, given that the islands are not inhabited and the reefs are already used regularly for recreational SCUBA diving.

The lower coral diversity communities B and C did not score highly on the various criteria assessed in Table 5.10, but nevertheless should not be excluded from conservation planning. In particular, the wave exposed S coast community C was not thoroughly surveyed because of large ocean swell (Figure 5.13). Many of these S coast reefs are highly prized for surfing, and as such draw large numbers of tourists to Bali each year. In the latter respect, their future conservation should be considered a priority for maintaining surf tourism on the island. Further offshore, some of these areas are crucial migration corridors for cetaceans and other species.

The presence of cool water upwelling and/or strong consistent current flow in some areas (eg. Nusa Penida, E Bali, as indeed also in Komodo and other areas of Indonesia) may be particularly important in buffering the incident reefs against rising sea temperatures associated with global climate change.

5.4.2 MPA network recommendations

In respect of establishing the MPA network, the following recommendations are made:

1. A multiple use MPA model, with different areas zoned for different levels of protection and use, is likely to be the most appropriate, given the broad range of activities that already occur on Bali's reefs. However, this model should include adequate core areas excluding extractive activities, to ensure conservation of key habitat and community types and foster replenishment.

2. As far as practicable, the MPA network should include representative and complementary areas encompassing the main coral community types (Figs. 7 and 12), and reefs of high conservation value (diversity, replenishment, rarity, Table 5.10).

3. As far as practicable, the network should include reefs subject to cool water upwelling and/or strong and consistent current flow, as a potential safeguard against increasing sea temperatures associated with global climate change over coming decades. Reefs of Nusa Penida and E Bali, particularly those under the influence of Lombok Strait, should be included in the network.

Figure 5.14. Reefs of high conservation priority, Bali, indicated by red stars.

4. There are many competing uses for Bali's coastal and marine resources, and it will be challenging to achieve the right balance among different levels of protection and use. Given the overwhelming importance of ocean-based tourism (surfing, diving and swimming), particular focus should be paid to maintaining healthy and attractive reef-scapes for these activities, and hence a focus on non-destructive, non-extractive activities in core zones.

5. Once an MPA network is established, enforcement of regulations will be crucial.

6. Consideration should be given to a 'User-pays' system (eg. Bunaken National Park) whereby visitors pay a nominal fee for access. This can provide significant funds for MPA management and benefits to local communities.

In respect of litter and water quality:

7. There is a widespread issue of litter and other forms of water pollution. A number of strategies may be employed / expanded to reduce the amount / impact of plastic and other pollution by: a) encouraging traditional packaging as much as practicable; b) continuing education campaigns in local mass media and schools; c) voluntary and funded litter clean-up activities on beaches and reefs.

8. Aim to improve stream and river water quality to reduce transport of litter / pollutants to reefs by restoring riparian vegetation; and with public education campaigns re inappropriate waste disposal.

ACKNOWLEDGEMENTS

We thank Dr. Mark Erdmann and staff of Conservation International Indonesia and CI International for organizing the survey. We also thank Dr. Joanne Wilson of The Nature Conservancy and Laure Katz of CI for help in the field, our colleagues from the Indonesian Institute of Sciences, the Indonesian Department of Nature Conservation, the Indonesian Ministry of Marine Affairs and Fisheries, and our other Indonesian and international colleagues for facilitating and supporting the field surveys. We also gratefully acknowledge Dr. Suharsono of the Indonesian Institute of Sciences, Dr. Mark Erdmann (CI), Mr. Erdi Lazuardi (CI Sorong Office), and Dr. Carden Wallace and staff of the Museum of Tropical Queensland (MTQ) for facilitating the continuing coral taxonomic studies. Dr. Charlie Veron (Coral Reef Research) and Dr. Carden Wallace (MTQ) provided much valued taxonomic advice.

REFERENCES

Abram N.J., M.K. Gagan, M.T. McCulloch, J. Chappell dan W.S. Hantoro, 2003. Coral reef death during the 1997 Indian Ocean Dipole linked to Indonesian wildfires. *Science* 301: 952.

Allen, G.R., 2007. Conservation hotspots of biodiversity and endemism for Indo-Pacific coral reef fishes. *Aquatic Conservation: Marine and Freshwater Ecosystems* 18: 541–556.

Allen, G. dan Steen, R. 1994. *Indo-Pacific Coral Reef Field Guide.* Singapore, Tropical Reef Research.

Barber, P.H., S.R. Palumbi, M.V. Erdmann dan M.K. Moosa, 2000. A marine Wallace's line? *Nature* 406: 392–693.

Barber, P.H., S.R. Palumbi, M.V. Erdmann dan M.K. Moosa, 2002. Sharp genetic breaks among populations of *Haptosquilla pulchella* (Stomatopoda) indicate limits to larval transport: patterns, causes, and consequences. *Molecular Ecology* 11: 659–674.

Barkman, J.J., H. Doing, dan Segal, S. 1964. Kritische bemerkungen und vorschlage zur quantitativen vegetationsanalyse. *Acta Botanica Neerlandica* 13: 394–419.

Colin, P.L. and Arneson, C. 1995. *Tropical Pacific Invertebrates.* Coral Reef Press, California, USA.

DeVantier, L.M., De'ath, G., Done, T.J. dan Turak, E. 1998. Ecological assessment of a complex natural system: a case-study from the Great Barrier Reef. *Ecological Applications* 8: 480–496.

DeVantier, L.M., De'ath, G., Klaus, R., Al-Moghrabi, S., Abdal-Aziz, M., Reinicke, G.B., dan Cheung, C.P.S. 2004. Reef-building corals and coral communities of the Socotra Islands, Yemen: A zoogeographic 'crossroads' in the Arabian Sea. *Fauna of Arabia* 20: 117–168.

DeVantier, L.M., Turak, E., dan Skelton, P. 2006. Ecological Assessment of the coral communities of Bunaken National Park: Indicators of management effectiveness. *Proceedings of the 10th International Coral Reef Symposium*, Okinawa.

DeVantier, L.M., Turak, E., dan Allen, G. 2008. Lesser Sunda Ecoregional Planning Coral Reef Stratification Reef- and Seascapes of the Lesser Sunda Ecoregion. Report to The Nature Conservancy, Jl. Pengembak No. 2, Sanur – Bali 80228, Indonesia, 30 hal. ditambah Lampiran.

Done, T.J. 1982. Patterns in the distribution of coral communities across the central Great Barrier Reef. *Coral Reefs* 1: 95–107.

Erdmann, M.V. dan R.B. Manning, 1998. Nine new stomatopod crustaceans from coral reef habitats in Indonesia and Australia. *Raffles Bulletin of Zoology* 46(2): 615–626.

Fukami, H., Chen, C.A., Budd, A.F., Collins, A., Wallace, C., Chuang, Y.-Y., Chen, C., Dai, C.-F., Iwao, K., Sheppard, C., dan Knowlton, N. 2008. Mitochondrial and nuclear genes suggest that stony coral sare monophyletic but most families of stony corals are not (Order Scleractinia, Class Anthozoa, Phylum Cnidaria). *PLOS One* http://dx.plos.org/10.1371/journal.pone.0003222.

Gosliner, T.M., Behrens, D.W. dan Williams, G.C. 1996. *Coral Reef Animals of the Indo-Pacific.* Monterey, USA. Sea Challengers.

Green A.L. dan P.J. Mous, 2007. Delineating the Coral Triangle, its ecoregions and functional seascapes. Report based on an expert workshop held at the TNC Coral Triangle Center, Bali Indonesia (April - May 2003), and subsequent consultations with experts held from 2005 to 2007. Version 4.0 (August 2007). Report from The Nature Conservancy, Coral Triangle Center (Bali, Indonesia) and the Global Marine Initiative, Indo-Pacific Resource Centre (Brisbane, Australia). 78 hal.

Hoeksema, B.W. 1989. Taxonomy, phylogeny and biogeography of mushroom corals (Scleractinia: Fungiidae). *Zoologische Verhandelingen* 254: 1–295.

Hoeksema, B.W. dan Putra, K.S. 2000. The reef coral fauna of Bali in the centre of marine biodiversity. *Proceedings of the 9th International Coral Reef Symposium*, Bali, Vol 1.

Hopley, D. 1982. *The Geomorphology of the Great Barrier Reef: Quaternary Development of Coral Reefs.* New York. John Wiley-Interscience, 453 hal.

Hopley, D., Parnell, K.E. dan Isdale, P.J. 1989. The Great Barrier Reef Marine Park: Dimensions and regional patterns. *Australian Geographic Studies* 27: 47–66.

Jongman, R.H.G., ter Braak, C.J.F. dan van Tongeren, O.F.R. 1995. Data analysis in community and landscape ecology. Cambridge University Press, 299 hal.

Miller, I.R. dan De'ath, G. 1995. Effects of training on observer performance in assessing benthic cover by means of the manta tow technique. *Marine and Freshwater Research* 47: 19–26.

Sheppard, C.R.C. dan Sheppard, A.L.S. 1991. Corals and coral communities of Arabia. *Fauna of Saudi Arabia* 12: 13–170.

Turak, E. 2002. *Assessment of coral biodiversity and coral reef health of the Snagihe-Talaud Islands, North Sulawesi, Indonesia, 2002.* Final Report to The Nature Conservancy.

Turak, E. 2004. *Coral Reef Surveys During TNC SEACMPA RAP of Wakatobi National Park, Southeast Sulawesi, Indonesia, May 2003.* Final Report to The Nature Conservancy.

Turak, E. 2005. *Coral Biodiversity and Reef Health.* Dalam: Mous, PJ, B. Wiryawan dan L.M. DeVantier (eds.) 2006. *Report on a rapid ecological assessment of Derawan Islands, Berau district, East Kalimantan, Indonesia, October 2003.* TNC Coastal Marin Program Report.

Turak, E. 2006a. *Corals and Coral Communities of the Komodo National Park.* Dalam: Beger, M dan Turak, E (2006) A Rapid Ecological Assessment of the reef fishes and scleractinian corals of Komodo National Park, Indonesia in 2005. The Nature Conservancy.

Turak, E. dan DeVantier, L.M. 2003. *Corals and coral communities of Bunaken National Park and nearby reefs, North Sulawesi*, Indonesia: Rapid ecological assessment of biodiversity and status. Final Report to the International Ocean Institute Regional centre for Australia and western Pacific.

Turak, E. dan DeVantier, L.M. 2009. Biodiversity and Conservation Priorities of Reef-building Corals in Nusa Penida. Final report to Conservation International, Indonesia.

Turak, E. dan DeVantier, L.M. 2011. *Field Guide to Reef-building Corals of Brunei Darussalam.* Department of Fisheries, Brunei Darussalam, 256 hal.

Turak, E. dan DeVantier, L. Dalam pencetakan. *Biodiversity and conservation priorities of reef-building corals in the Papuan Bird's Head Seascape.* Conservation International, Indonesia.

Turak, E. dan Shouhoka, J. 2003. *Coral diversity and status of the coral reefs in the Raja Ampat islands, Papua province, Indonesia, November 2002.* Final Report to The Nature Conservancy

Turak, E., Wakeford, M. dan Done, T.J. 2003. *Kepulauan Banda rapid ecological assessment, May 2002: Assessment of coral biodiversity and coral reef health.* Dalam, Mous, P.J (ed), Report on a rapid ecological assessment of the Kepulauan Banda, Maluku, Eastern Indonesia, held April 28 – May *5 2002,* TNC and UNESCO publication, 150 hal.

van der Maarel, E. 1979. Transformation of cover-abundance values in phytosociology and its effects on community similarity. *Vegetatio* 39: 97–114.

van Woesik, R. 1997. A comparative survey of coral reefs in south-eastern Bali, Indonesia, 1992 and 1997. Laporan tidak dipublikasi.

van Woesik, R. 2004. Comment on "Coral Reef Death During the 1997 Indian Ocean Dipole Linked to Indonesian Wildfires". Science 303: 1297.

Veron, J.E.N., DeVantier, L.M., Turak, E., Green, A.L., Kininmonth, S., Allen, G.R., Stafford-Smith, M.G., Mous, P.A. dan Petersen, N.A. (tidak dipublikasi) Global coral biodiversity: a blueprint for reef conservation.

Veron, J.E.N. 1986. *Corals of Australia and the Indo-Pacific.* Angus and Robertson, Australia, 644 hal.

Veron, J.E.N. 1990. New Scleractinia from Japan and other Indo-west Pacific countries. *Galaxea* 9: 95–173.

Veron, J.E.N. 1993. *A Biogeographic Database of Hermatypic Corals Species of the Central Indo-Pacific Genera of the World.* Australian Institute of Marine Science Monograph Series Vol. 10, 433 hal.

Veron, J.E.N. 1995. *Corals in Space and Time The Biogeography and Evolution of the Scleractinia.* University of New South Wales Press, 321 hal.

Veron, J.E.N. 1998. *Corals of the Milne Bay Region of Papua New Guinea.* Dalam: Werner, TA dan Allen GR (eds). A *rapid biodiversity assessment of the coral reefs of Milne Bay*

Province, Papua New Guinea. Conservation International, RAP Working Papers, 11.

Veron, J.E.N. 2000. *Corals of the World.* Australian Institute of Marine Science publ.

Veron, J.E.N. 2002. *New Species Described in Corals of the World.* Australian Institute of Marine Science Monograph Series, Vol. 11. Australian Institute of Marine Science publ.

Veron, J.E.N. dan Pichon, M. 1976. Scleractinia of Eastern Australia. Part I Families Thamnasteriidae, Astrocoeniidae, Pocilloporidae. Australian National University Press, Canberra, Australian Institute of Marine Science Monograph Series 1, 86 hal.

Veron, J.E.N. dan Pichon, M. 1980. Scleractinia of Eastern Australia. Part III Families Agariciidae, Siderastreidae, Fungiidae, Oculinidae, Merulinidae, Mussidae, Pectiniidae, Caryophylliidae, Dendrophylliidae. Australian National University Press, Canberra, Australian Institute of Marine Science Monograph Series 4, 422 hal.

Veron, J.E.N. dan Pichon, M. 1982. Scleractinia of Eastern Australia. Part IV. Family Poritidae Australian National University Press, Canberra, Australian Institute of Marine Science Monograph Series 5, 159 hal.

Veron, J.E.N., Pichon, M. dan Wijsman-Best, M. 1977. Scleractinia of Eastern Australia. Part II Families Faviidae, Trachyphylliidae. Australian National University Press, Canberra, Australian Institute of Marine Science Monograph Series 1, 233 hal.

Veron, J.E.N. dan Wallace, C.C. 1984. Scleractinia of Eastern Australia. Part V Family Acroporidae. Australian National University Press, Canberra, Australian Institute of Marine Science Monograph Series 1, 485 hal.

Veron, J.E.N., DeVantier, L.M., Turak, E., Green, A.L., Kininmonth, S., dan Petersen, N.A. 2009. Delineating the Coral Triangle. *Galaxea* 11: 91–100.

Wallace, C.C. 1999. *Staghorn corals of the World.* CSIRO publ., Australia.

Wallace, C.C. dan Wolstenholme, J. 1998. Revision of the coral genus *Acropora* (Scleractinia: Astrocoeniina: Acroporidae) in Indonesia. *Zoological Journal of the* Linnean Society 123: 199–384.

Wallace, C.C., Turak, E. dan DeVantier, L.M. Submitted. Novelty, parallelism and record site diversity in a conservative coral genus: three new species of *Astreopora* (Scleractinia; Acroporidae) from the Papuan Bird's Head Seascape. *Proc. Royal Society B.*

World Fish Center (diakses pada 19 Mei 2007). An Institutional Analysis of Sasi Laut in Maluku, Indonesia. http://www.worldfishcenter.org/Pubs/Sasi/.pdf

Appendix 5.1. Characteristics of survey stations. Nusa Penida, November 2008 and Bali, April–May 2011. EXP – Exposure rank; RD – Reef Development rank; VIS – Underwater Visibility (water clarity, in meters); WT – Water Temperature (degrees centigrade, see Methods).

Location	Station name	Station	Date	Latitude, S	Longitude, E	EXP	RD	VIS	SP
Lembongan	Lembongan Bay Pantoon	**1.2**	20-Nov-08	8°40.455	115°26.328	3	4	20	23
Lembongan	Mushroom Bay North	**2.2**	20-Nov-08	8°40.781	115°25.977	3	4	20	23
Nusa Penida	Toya Pakeh	**3.1**	21-Nov-08	8°40.997	115°28.957	2	4	25	29
Nusa Penida	Toya Pakeh	**3.2**	21-Nov-08	8°39.84	115°28.017	3	4	20	29
Lembongan	Mangrove N Lembongan	**4.1**	21-Nov-08	8°39.84	115°28.017	2	4	20	28
Lembongan	Mangrove N Lembongan	**4.2**	21-Nov-08	8°47.943	115°31.584	3	4	20	29
Nusa Penida	Manta Point	**5.1**	22-Nov-08	8°47.943	115°31.584	3	1	15	26
Nusa Penida	Old Manta Bay	**6.1**	22-Nov-08	8°45.242	115°28.194	3	1	12	28
Nusa Penida	Crystal Bay South	**7.1**	26-Nov-08	8°42.977	115°27.431	2	3	25	27
Nusa Penida	Crystal Bay South	**7.2**	22-Nov-08	8°42.977	115°27.431	3	3	25	29
Nusa Penida	Batu Abah	**8.1**	23-Nov-08	8°46.461	115°37.616	2	2	30	28
Nusa Penida	Batu Abah	**8.2**	23-Nov-08	8°46.461	115°37.616	3	2	25	29
Nusa Penida	South of Batu Abah	**9.1**	23-Nov-08	8°47.848	115°36.409	2	2	10	28
Nusa Penida	South of Batu Abah	**9.2**	23-Nov-08	8°47.848	115°36.409	3	2	10	28
Nusa Penida	Malibu Point	**10.1**	24-Nov-08	8°42.833	115°35.623	2	4	20	29
Nusa Penida	Malibu Point	**10.2**	24-Nov-08	8°42.833	115°35.623	3	4	5	30
Nusa Penida	Batunggul	**11.1**	24-Nov-08	8°41.381	115°34.923	2	3	30	29
Nusa Penida	Batunggul	**11.2**	24-Nov-08	8°41.381	115°34.923	3	3	20	29
Nusa Penida	Buyuk	**12.1**	25-Nov-08	8°40.47	115°32.596	2	3	25	29
Nusa Penida	Buyuk	**12.2**	25-Nov-08	8°40.47	115°32.596	3	3	10	29
Nusa Penida	Sental	**13.1**	27-Nov-08	8°40.576	115°31.691	2	3	20	28
Nusa Penida	Sental	**13.2**	25-Nov-08	8°40.576	115°31.691	3	3	15	29
Lembongan	Ceningen channel	**14.1**	27-Nov-08	8°41.079	115°27.942	2	4	20	28
Lembongan	Ceningen channel	**14.2**	26-Nov-08	8°41.079	115°27.942	2	4	15	29
Lembongan	Mushroom Bay South	**15.2**	26-Nov-08	8°40.763	115°25.852	3	2	25	27
Nusa Penida	Crystal Bay Rock	**16.1**	29-Nov-08	8°42.905	115°27.338	2	2	20	28
Nusa Penida	Crystal Bay Rock	**16.2**	27-Nov-08	8°42.905	115°27.338	3	2	20	28
Nusa Penida	Sekolah Dasar	**17.1**	28-Nov-08	8°40.349	115°30.515	2	4	25	27
Nusa Penida	Sekolah Dasar	**17.2**	28-Nov-08	8°40.349	115°30.515	3	4	25	27
Bali SE	Terora, Sanur	**1.1**	29-Apr-11	8°46.228	115°13.805	3	4	8	29
Sanur	Terora, Sanur	**1.2**	29-Apr-11	8°46.228	115°13.805	4	4	6	29
Nusa Dua	Glady Willis, Nusa Dua	**2.1**	29-Apr-11	8°41.057	115°16.095	3	4	6	29
	Glady Willis, Nusa Dua	**2.2**	29-Apr-11	8°41.057	115°16.095	3	4	8	29
Sanur	Sanur Channel N side	**3.1**	29-Apr-11	8°42.625	115°16.282	2	4	8	29
Sanur	Sanur Channel	**3.2**	29-Apr-11	8°42.625	115°16.282	4	4	4	28
Nusa Dua	Kutuh Temple	**4.1**	30-Apr-11	8°50.617	115°12.336	4	4	6	28
	Nusa Dua - Public beach	**5.1**	30-Apr-11	8°50.617	115°12.336	3	4	12	29
Nusa Dua	Nusa Dua - Public beach	**5.2**	30-Apr-11	8°48.025	115°14.356	4	4	10	28
Nusa Dua	Melia Bali hotel	**6.1**	30-Apr-11	8°47.608	115°14.192	3	4	10	28
	Melia Bali hotel	**6.2**	30-Apr-11	8°47.608	115°14.192	2	4	8	29
Padang Bai	West Gili Mimpang (Batu Tiga)	**7.1**	1-Mei-11	8°31.527	115°34.519	1	2	20	29
Padang Bai	West Gili Mimpang (Batu Tiga)	**7.2**	1-Mei-11	8°31.527	115°34.519	3	2	20	28

table continued on next page

Appendix 5.1. *continued.*

Location	Station name	Station	Date	Latitude, S	Longitude, E	EXP	RD	VIS	SP
Padang Bai	East Gili Mimpang (Batu Tiga)	**8.1**	1-Mei-11	8°31.633	115°34.585	2	2	20	28
Padang Bai	East Gili Mimpang (Batu Tiga)	**8.2**	1-Mei-11	8°31.633	115°34.585	2	2	20	29
Padang Bai	Jepun, Amuk Bay, Candi Dasa	**9.1**	1-Mei-11	8°31.138	115°34.619	1	4	7	29
Padang Bai	Jepun, Amuk Bay, Candi Dasa	**9.2**	1-Mei-11	8°31.138	115°34.619	2	4	6	28
Padang Bai	Gili Tepekong, Candi Dasa	**10.1**	2-Mei-11	8°31.885	115°35.167	2	2	30	28
Padang Bai	Gili Tepekong, Candi Dasa	**10.2**	2-Mei-11	8°31.885	115°35.167	2	2	25	29
Padang Bai	Gili Biaha/ Tanjung Pasir Putih	**11.1**	2-Mei-11	8°30.27	115°36.771	1	2	15	29
Padang Bai	Gili Biaha/Tanjung Pasir Putih	**11.2**	2-Mei-11	8°30.27	115°36.771	3	2	15	28
NE Bali	Seraya	**12.1**	3-Mei-11	8°26.01	115°41.274	3	1	6	28
NE Bali	Seraya	**12.2**	3-Mei-11	8°26.01	115°41.274	2	1	10	29
NE Bali	Gili Selang North	**13.1**	3-Mei-11	8°23.841	115°42.647	1	3	25	29
NE Bali	Gili Selang North	**13.2**	3-Mei-11	8°23.841	115°42.647	3	1	16	28
NE Bali	Gili Selang South	**14.1**	3-Mei-11	8°24.079	115°42.679	3	1	12	29
NE Bali	Gili Selang South	**14.2**	3-Mei-11	8°24.079	115°42.679	2	2	20	29
NE Bali	Bunutan, Amed	**15.1**	4-Mei-11	8°20.731	115°40.826	1	1	20	30
NE Bali	Bunutan, Amed	**15.2**	4-Mei-11	8°20.731	115°40.826	3	2	20	30
NE Bali	Jemeluk, Amed	**16.1**	4-Mei-11	8°20.221	115°39.617	2	3	20	30
NE Bali	Jemeluk, Amed	**16.2**	4-Mei-11	8°20.221	115°39.617	1	3	20	30
NE Bali	Kepa, Amed	**17.1**	4-Mei-11	8°20.024	115°39.244	1	3	20	30
NE Bali	Kepa, Amed	**17.2**	4-Mei-11	8°20.024	115°39.244	3	3	20	30
NE Bali	Batu Kelibit, Tulamben	**18.1**	5-Mei-11	8°16.696	115°35.826	2	2	20	30
NE Bali	Batu Kelibit, Tulamben	**18.2**	5-May-11	8°16.696	115°35.826	2	2	20	30
NE Bali	Tukad Abu, Tulamben	**19.1**	5-May-11	8°17.603	115°36.599	1	1	15	30
NE Bali	Tukad Abu, Tulamben	**19.2**	5-May-11	8°17.603	115°36.599	3	2	10	30
NE Bali	Gretek	**20.1**	6-May-11	8°8.969	115°24.733	2	2	3	28
NE Bali	Gretek	**20.2**	6-May-11	8°8.969	115°24.733	2	2	5	30
NE Bali	Penutukang	**21.1**	6-May-11	8°8.27	115°23.622	2	2	6	29
NE Bali	Penutukang	**21.2**	6-May-11	8°8.27	115°23.622	2	2	5	30
NW Bali	Puri Jati	**22**	7-May-11	8°11.032	114°54.869	2	1	6	29
NW Bali	Kalang Anyar	**23**	7-May-11	8°11.344	114°53.841	2	1	4	29
NW Bali	Taka Pemutaran	**24.1**	8-May-11	8°7.775	114°40.007	2	2	20	29
NW Bali	Taka Pemutaran	**24.2**	8-May-11	8°7.775	114°40.007	3	2	16	29
NW Bali	Sumber Kima	**25.1**	8-May-11	8°6.711	114°36.451	2	4	15	29
NW Bali	Sumber Kima	**25.2**	8-May-11	8°6.711	114°36.451	3	4	12	29
NW Bali	Menjangan North	**26.1**	9-May-11	8°5.467	114°30.131	2	4	25	30
NW Bali	Menjangan North	**26.2**	9-May-11	8°5.467	114°31.131	3	4	18	30
NW Bali	Menjangan East	**28.1**	9-May-11	8°5.813	114°31.608	2	3	16	28
NW Bali	Menjangan East	**28.2**	9-May-11	8°5.813	114°31.608	3	3	10	30
NW Bali	Secret Bay, Muck dive	**29**	10-May-11	8°9.862	114°26.302	1	1	4	28
NW Bali	Secret Bay, Reef north shore	**30**	10-May-11	8°9.771	114°27.116	2	4	6	28
NW Bali	Pearl farm	**31.1**	11-May-11	8°13.911	114°27.249	2	3	3	28
NW Bali	Pearl farm	**31.2**	11-May-11	8°13.911	114°27.249	3	3	3	28
NW Bali	Pearl farm	**32.2**	11-May-11	8°14	114°27.463	2	1	4	29

Appendix 5.2. Visual estimates of percent cover of sessile benthic attributes and substratum types, and depth and station tallies for hermatypic coral species richness, Nusa Penida, November 2008 and Bali, April–May 2011. max - maximum depth (m); min – minimum depth (m). Sessile Benthos: HS – Hard Substrate; HC – Hard Coral; SC – Soft Coral; MA – Macro-Algae; TA – Turf Algae; CA – Coralline Algae; DC – recently Dead Coral; AD – All Dead coral. Substratum types: CP – Continuous Pavement; LB – Large Blocks (>2m diam.); SB – Small Blocks (<2m diam.); RBL – Rubble; SN – Sand.

Station name	Site	Max	Min	Slope	HS	HC	SC	MA	TA	CA	DC	AD	CP	LB	SB	RBL	SN	No. of species	Station total
Lembongan Bay Pantoon	1.2	13	5	5	95	60	10	20	5	5	1	1	80	10	5	0	5	81	81
Mushroom Bay North	2.2	6,5	2	3	90	50	5	20	5	2	0	1	80	5	5	0	10	74	74
Toya Pakeh	3.1	23	10	20	85	60	10	0	5	10	1	0	70	10	5	10	5	79	
Toya Pakeh	3.2	8	1	3	80	50	30	0	10	10	0	2	60	15	5	15	5	79	114
Mangrove N Lembongan	4.1	27	10	20	100	40	5	0	5	5	0	0	85	10	5	0	0	88	
Mangrove N Lembongan	4.2	8	1	10	80	50	10	0	5	5	0	0	70	5	5	10	10	90	134
Manta Point	5.1	34	10	10	90	10	5	0	20	0	0	0	100	0	0	0	0	70	70
Old Manta Bay	6.1	30	12	5	100	3	20	15	20	0	0	0	100	0	0	0	0	42	42
Crystal Bay South	7.1	29	10	30	70	50	20	0	5	10	0	0	65	0	5	25	5	52	
Crystal Bay South	7.2	8	1	5	90	60	30	5	5	5	1	2	70	15	5	5	5	96	123
Batu Abah	8.1	35	10	20	90	50	2	0	5	10	1	3	80	5	5	5	5	89	
Batu Abah	8.2	8	1,5	5	95	50	10	0	5	10	1	3	85	5	5	5	0	76	121
South of Batu Abah	9.1	29	10	10	85	15	5	0	20	5	1	3	55	20	10	10	5	80	
South of Batu Abah	9.2	8	1,5	5	90	20	5	0	20	10	1	2	50	30	10	5	5	67	116
Malibu Point	10.1	40	10	30	90	30	5	0	5	10	1	5	80	5	5	10	0	90	
Malibu Point	10.2	8	1	5	90	30	1	0	20	5	1	3	60	20	10	5	5	101	141
Batunggul	11.1	38	10	20	95	20	2	0	5	5	1	2	70	20	5	0	5	92	
Batunggul	11.2	8	1	10	95	50	0	0	20	10	0	0	70	20	5	5	0	95	140
Buyuk	12.1	38	10	20	95	30	30	0	5	5	0	0	80	10	5	0	5	62	
Buyuk	12.2	8	1	10	80	30	40	0	10	5	0	0	65	5	10	5	15	78	115
Sental	13.1	38	10	30	80	20	10	0	10	5	0	0	60	10	10	10	10	88	
Sental	13.2	8	1	5	70	20	30	0	10	5	1	3	50	10	10	20	10	72	126
Ceningen channel	14.1	31	10	10	70	20	10	0	10	5	1	3	55	10	5	20	10	73	
Ceningen channel	14.2	8	1	5	60	20	20	2	10	5	1	2	40	10	10	10	30	78	119
Mushroom Bay South	15.2	10	3	3	60	30	10	5	5	10	1	3	40	15	5	20	20	81	81
Crystal Bay Rock	16.1	45	10	30	90	20	10	0	5	10	0	0	75	0	5	5	5	82	
Crystal Bay Rock	16.2	10	2	5	90	30	20	0	5	10	0	0	80	5	5	5	5	61	103
Sekolah Dasar	17.1	38	10	20	80	30	3	0	0	5	0	0	70	5	5	0	20	73	
Sekolah Dasar	17.2	8	1	5	90	60	5	0	5	10	0	0	70	15	5	5	5	103	138
Terora, Sanur	1.1	13	6	20	90	15	5	1	20	20	1	5	50	20	20	3	7	880	
Terora, Sanur	1.2	6	2	2	100	30	20	5	10	10	0	0	90	5	5	0	0	83	126
Glady Willis, Nusa Dua	2.1	10	5	20	80	20	5	0	10	5	0	0	60	10	10	5	15	88	
Glady Willis, Nusa Dua	2.2	5	0,5	10	95	25	5	2	20	15	1	2	70	15	10	0	5	90	133
Sanur Channel N side	3.1	15	7	40	90	10	5	2	10	30	1	3	70	10	10	5	5	57	
Sanur Channel	3.2	6	2	2	100	30	5	0	10	10	0	0	90	5	5	0	0	44	79
Kutuh Temple	4.1	13	8	5	80	10	30	10	0	10	0	0	80	0	0	0	20	62	62
Nusa Dua - Public beach	5.1	16	7	30	95	10	20	10	10	20	1	2	85	10	0	0	5	67	
Nusa Dua - Public beach	5.2	7	2	2	100	10	5	5	20	10	0	0	90	10	0	0	0	65	102

table continued on next page

Appendix 5.2. *continued.*

Station name	Site	Max	Min	Slope	HS	HC	SC	MA	TA	CA	DC	AD	CP	LB	SB	RBL	SN	No. of species	Station total
Melia Bali hotel	6.1	15	7	5	90	30	10	5	5	10	0	0	80	5	5	0	10	66	
Melia Bali hotel	6.2	7	2	15	90	25	20	5	20	20	1	3	70	10	10	5	5	95	121
West Gili mimpang (Batu Tiga)	7.1	23	9	10	50	15	5	2	20	30	1	25	20	20	10	30	20	100	
West Gili Mimpang (Batu Tiga)	7.2	8	4	5	70	40	5	0	5	10	0	0	50	10	10	10	20	82	142
East Gili Mimpang (Batu Tiga)	8.1	30	10	30	70	30	5	1	10	10	0	0	50	10	10	20	10	84	
East Gili Mimpang (Batu Tiga)	8.2	9	5	20	90	35	5	2	20	20	1	5	50	20	20	5	5	79	122
Jepun, Amuk Bay, Candi Dasa	9.1	21	9	30	40	15	2	5	40	10	3	20	20	10	10	30	30	82	
Jepun, Amuk Bay, Candi Dasa	9.2	8	1	10	30	10	2	0	30	5	5	20	20	5	5	60	10	87	126
Gili Tepekong, Candi Dasa	10.1	33	11	20	70	30	3	0	5	10	0	0	50	10	10	10	20	99	
Gili Tepekong, Candi Dasa	10.2	10	3	30	100	50	5	1	10	10	1	3	70	30	0	0	0	83	137
Gili Biaha/ Tanjung Pasir Putih	11.1	24	9	10	50	15	3	1	30	20	1	10	10	20	20	20	30	108	
Gili Biaha/ Tanjung Pasir Putih	11.2	8	1	20	80	20	2	0	5	10	0	0	60	10	10	20	0	76	142
Seraya	12.1	16	10	5	20	30	30	0	10	5	0	0	0	10	10	0	80	67	
Seraya	12.2	8	3	10	80	30	40	1	10	10	1	2	0	50	30	0	20	79	110
Gili Selang North	13.1	31	9	25	50	15	15	1	10	10	1	2	20	20	10	10	40	78	
Gili Selang North	13.2	8	1	2	95	40	30	0	5	10	0	0	40	30	25	0	5	76	117
Gili Selang South	14.1	31	10	30	70	30	10	0	10	10	0	0	40	20	10	0	30	72	
Gili Selang South	14.2	9	3	15	90	35	15	2	20	20	1	2	50	20	20	5	5	92	125
Bunutan, Amed	15.1	32	9	20	50	5	5	1	10	10	1	2	10	20	20	20	30	46	
Bunutan, Amed	15.2	8	1	5	90	60	0	0	19	10	0	0	50	30	10	0	10	97	120
Jemeluk, Amed	16.1	31	10	40	20	30	0	0	30	10	0	0	0	10	10	80	0	104	
Jemeluk, Amed	16.2	8	1	10	80	35	5	3	20	20	1	10	50	20	10	10	10	132	181
Kepa, Amed	17.1	30	9	15	50	15	3	1	30	20	1	3	20	20	10	10	20	111	
Kepa, Amed	17.2	8	1	2	80	30	1	0	20	5	0	0	50	10	20	10	5	94	158
Batu Kelibit, Tulamben	18.1	35	10	60	100	40	0	0	10	5	0	0	80	10	10	0	0	117	
Batu Kelibit, Tulamben	18.2	9	1	15	95	20	2	1	30	20	1	3	70	10	15	2	3	95	157
Tukad Abu, Tulamben	19.1	33	9	25	10	2	5	1	5	10	1	2	0	5	5	10	40	68	
Tukad Abu, Tulamben	19.2	8	2	10	90	40	1	0	20	10	0	0	60	10	20	5	5	121	156
Gretek	20.1	24	10	20	40	20	1	0	20	10	10	5	10	20	10	0	60	80	
Gretek	20.2	9	2	10	90	20	3	5	30	10	1	5	60	20	10	5	5	121	150
Penuktukan	21.1	25	10	20	40	20	0	0	20	0	0	1	10	20	10	0	60	76	
Penuktukan	21.2	9	2	30	90	35	2	2	30	10	1	10	60	20	10	5	5	132	164
Puri Jati	22	26	1	10	0	2	0	0	0	0	0	0	0	0	0	0	60	2	2

table continued on next page

Appendix 5.2. *continued.*

Station name	Site	Max	Min	Slope	HS	HC	SC	MA	TA	CA	DC	AD	CP	LB	SB	RBL	SN	No. of species	Station total
Kalang Anyar	23	15	1	5	2	<1	0	0	0	0	0	0	0	0	2	0	40	8	8
Taka Pemutaran	24.1	35	10	30	70	20	5	0	10	5	1	5	50	10	10	10	20	90	
Taka Pemutaran	24.2	8	3	2	80	30	3	0	10	10	1	3	50	10	20	10	10	97	138
Sumber Kima	25.1	34	10	60	95	30	5	1	10	10	0	0	80	10	5	5	0	104	
Sumber Kima	25.2	8	1	5	80	30	5	1	10	5	0	0	50	10	20	10	10	109	154
Menjangan North	26.1	39	10	40	90	30	3	0	5	10	0	0	80	5	5	5	5	115	
Menjangan North	26.2	8	1	2	70	70	3	0	5	5	0	0	60	0	10	0	30	106	168
Menjangan East	28.1	38	10	90	100	20	10	0	10	10	0	0	100	0	0	0	0	82	
Menjangan East	28.2	8	1	20	95	20	40	0	20	5	0	0	90	0	5	0	5	111	150
Secret Bay, Muck dive	29	8	1	10	5	3	0	0	2	0	0	0	0	0	5	0	95	21	21
Secret Bay, Reef north shore	30	13	2	5	70	60	0	2	10	0	0	0	60	0	10	10	20	44	44
Pearl farm	31.1	21	10	20	80	20	20	0	20	0	0	5	70	5	5	10	10	47	
Pearl farm	31.2	8	2	10	90	20	20	0	20	0	0	5	60	20	10	0	10	48	75
Pearl farm	32.2	12	2	5	80	10	10	0	20	0	0	0	40	20	20	0	20	45	45

Appendix 5.3. Coral species check-list for Bali an adjacent regions, including Komodo, Wakatobi. Derewan and Bunaken NP. Species' records for each location have been updated with continuing taxonomic study. • – confirmed species; U – unconfirmed, based on observational and/or photographic evidence, and requiring confirmation; H – Hoeksema & Putra, 2000; 1998. KOM – Komodo, (Turak, 2006); WAK – Wakatobi, (Turak, 2004); BNP – Bunaken NP (DeVantier et al. 2006) and DER – Derewan (Turak, 2005).

Zooxanthellate scleractinia	BALI	KOM	WAK	BNP	DER
Family Astrocoeniidae Koby, 1890					
Genus *Stylocoeniella* Yabe and Sugiyama, 1935					
Stylocoeniella armata (Ehrenberg, 1834)	•	•	•	•	•
Stylocoeniella guentheri Bassett-Smith, 1890	•	•	•	•	•
Genus *Palauastrea* Yabe and Sugiyama, 1941			•		
Palauastrea ramosa Yabe and Sugiyama, 1941	•	•		•	•
Genus *Madracis* Milne Edwards and Haime, 1849					
Madracis kirbyi Veron and Pichon, 1976	•				•
Family Pocilloporidae Gray, 1842					
Genus *Pocillopora* Lamarck, 1816					
Pocillopora ankeli Scheer and Pillai, 1974	•	•	•		•
Pocillopora damicornis (Linnaeus, 1758)	•	•	•	•	•
Pocillopora danae Verrill, 1864	•	•	•	•	•
Pocillopora elegans Dana, 1846	•				•
Pocillopora eydouxi Milne Edwards and Haime, 1860	•	•	•	•	•
Pocillopora kelleheri Veron, 2002	•		•	•	•
Pocillopora meandrina Dana, 1846	•	•	•	•	•
Pocillopora verrucosa (Ellis and Solander, 1786)	•	•	•	•	•
Pocillopora woodjonesi Vaughan, 1918		•		•	•
Genus *Seriatopora* Lamarck, 1816					
Seriatopora aculeata Quelch, 1886	•	•	•	•	•
Seriatopora caliendrum Ehrenberg, 1834	•	•	•	•	•
Seriatopora dendritica Veron, 2002				•	•
Seriatopora guttatus Veron, 2002	•		•		•
Seriatopora hystrix Dana, 1846	•	•	•	•	•
Seriatopora stellata Quelch, 1886			•	•	•
Genus *Stylophora* Schweigger, 1819					
Stylophora pistillata Esper, 1797	•	•	•	•	•
Stylophora subseriata (Ehrenberg, 1834)	•	•	•	•	•
Family Acroporidae Verrill, 1902					
Genus *Montipora* Blainville, 1830					
Montipora aequituberculata Bernard, 1897	•	•	•	•	•
Montipora altasepta Nemenzo, 1967	•	•		•	•
Montipora angulata (Lamarck, 1816)	•				•
Montipora cactus Bernard, 1897		•	•	•	•
Montipora calcarea Bernard, 1897	•	•	•	•	•
Montipora caliculata (Dana, 1846)	•	•	•	•	•
Montipora capitata Dana, 1846	•	•	•	•	•
Montipora capricornis Veron, 1985	•				
Montipora cebuensis Nemenzo, 1976	•	•	•	•	•
Montipora confusa Nemenzo, 1967	•	•	•	•	•

table continued on next page

Appendix 5.3. *continued.*

Zooxanthellate scleractinia	BALI	KOM	WAK	BNP	DER
Montipora corbettensis Veron and Wallace, 1984	•	•	•	•	•
Montipora crassituberculata Bernard, 1897	•	•			•
Montipora danae (Milne Edwards and Haime, 1851)	•	•	•	•	•
Montipora deliculata Veron, 2002	•		•	•	•
Montipora digitata (Dana, 1846)	•	•	•	•	•
Montipora dilatata Studer, 1901		•			
Montipora efflorescens Bernard, 1897	•	•	•	•	•
Montipora effusa Dana, 1846	•				
Montipora florida Nemenzo, 1967	U	•	•	•	•
Montipora floweri Wells, 1954	•	•	•	•	•
Montipora foliosa (Pallas, 1766)	•	•	•	•	•
Montipora foveolata (Dana, 1846)	•	•	•	•	•
Montipora friabilis Bernard, 1897	•	•		•	•
Montipora gaimardi Bernard, 1897	•				
Montipora grisea Bernard, 1897	•	•	•	•	•
Montipora hirsuta Nemenzo, 1967	•				
Montipora hispida (Dana, 1846)	•	•	•	•	•
Montipora hodgsoni Veron, 2002	•		•	•	•
Montipora hoffmeisteri Wells, 1954	•	•	•	•	•
Montipora incrassata (Dana, 1846)	•	•	•	•	•
Montipora informis Bernard, 1897	•	•	•	•	•
Montipora mactanensis Nemenzo, 1979	•	•	•	•	•
Montipora malampaya Nemenzo, 1967				•	•
Montipora millepora Crossland, 1952	•	•	•	•	•
Montipora mollis Bernard, 1897	•	•		•	•
Montipora monasteriata (Forskål, 1775)	•	•	•	•	•
Montipora nodosa (Dana, 1846)	•	•	•	•	•
Montipora palawanensis Veron, 2002	•	•	•	•	•
Montipora peltiformis Bernard, 1897	•		•	•	
Montipora porites Veron, 2002	•		•	•	
Montipora samarensis Nemenzo, 1967	•	•	•	•	•
Montipora spongiosa (Ehrenberg, 1834)		•			
Montipora spongodes Bernard, 1897	•		•	•	•
Montipora spumosa (Lamarck, 1816)	•		•	•	•
Montipora stellata Bernard, 1897	•	•		•	•
Montipora tuberculosa (Lamarck, 1816)	•	•	•	•	•
Montipora turgescens Bernard, 1897	•	•	•	•	•
Montipora turtlensis Veron dan Wallace, 1984	•	•	•		•
Montipora undata Bernard, 1897	•	•	•	•	•
Montipora venosa (Ehrenberg, 1834)	•			•	•
Montipora verrucosa (Lamarck, 1816)	•	•		•	•
Montipora verruculosus Veron, 2002		•	•	•	•
Montipora vietnamensis Veron, 2002	•	•	•	•	•

table continued on next page

Appendix 5.3. *continued.*

Zooxanthellate scleractinia	BALI	KOM	WAK	BNP	DER
Genus *Anacropora* Ridley, 1884					
Anacropora forbesi Ridley, 1884	•	•		•	•
Anacropora matthai Pillai, 1973	•				•
Anacropora puertogalerae Nemenzo, 1964	•	•	•	•	•
Anacropora reticulate Veron dan Wallace, 1984	•	•	•	•	•
Anacropora spinosa Rehberg, 1892				•	•
Genus *Acropora* Oken, 1815					
Acropora abrolhosensis Veron, 1985			•	•	•
Acropora abrotanoides (Lamarck, 1816)	•	•	•	•	•
Acropora aculeus (Dana, 1846)	•	•	•	•	•
Acropora acuminata (Verril, 1864)	•	•	•	•	•
Acropora anthocercis (Brook, 1893)	•	•	•	•	•
Acropora aspera (Dana, 1846)	•	•		•	•
Acropora austera (Dana, 1846)	•	•	•	•	•
Acropora awi Wallace dan Wolstenholme, 1998	•				•
Acropora bifurcate Nemenzo, 1971					•
Acropora carduus (Dana, 1846)		•	•	•	•
Acropora caroliniana Nemenzo, 1976			•	•	•
Acropora cerealis (Dana, 1846)	•	•	•	•	•
Acropora clathrata (Brook, 1891)	•	•	•	•	•
Acropora convexa (Dana, 1846)				•	
Acropora cophodactyla (Brook, 1892)	U		•	•	•
Acropora copiosa Nemenzo, 1967	U		•		•
Acropora cytherea (Dana, 1846)	•	•	•	•	•
Acropora derawanensis Wallace, 1997			•		•
Acropora desalwii Wallace, 1994			•	•	
Acropora digitifera (Dana, 1846)	•	•	•	•	•
Acropora divaricata (Dana, 1846)	•	•	•	•	•
Acropora donei Veron dan Wallace, 1984	•	•	•	•	•
Acropora echinata (Dana, 1846)	•		•	•	•
Acropora efflorescens (Dana, 1846)	•				
Acropora elegans Milne Edwards dan Haime, 1860	•		•		•
Acropora elseyi (Brook, 1892)	•	•	•		•
Acropora florida (Dana, 1846)	•	•	•	•	•
Acropora formosa (Dana, 1846)	•	•	•	•	•
Acropora gemmifera (Brook, 1892)	•	•	•	•	•
Acropora glauca (Brook, 1893)	•				
Acropora grandis (Brook, 1892)		•	•	•	•
Acropora granulosa (Milne Edwards dan Haime, 1860)	•	•	•	•	•
Acropora halmareae Wallace & Wolstenholme, 1998			•		
Acropora hoeksemai Wallace, 1997	U		•	•	•
Acropora horrida (Dana, 1846)	•	•	•	•	•
Acropora humilis (Dana, 1846)	•	•	•	•	•

table continued on next page

Appendix 5.3. *continued.*

Zooxanthellate scleractinia	BALI	KOM	WAK	BNP	DER
Acropora hyacinthus (Dana, 1846)	•	•	•	•	•
Acropora indonesia Wallace, 1997	•	•		•	•
Acropora insignis Nemenzo, 1967	•	•	•	•	•
Acropora jacquelineae Wallace, 1994			•		
Acropora kimbeensis Wallace, 1991	•				
Acropora kirstyae Veron dan Wallace, 1984	•		•	•	•
Acropora latistella (Brook, 1891)	•	•	•	•	•
Acropora listeri (Brook, 1893)	•	•	•	•	•
Acropora loisetteae Wallace, 1994					•
Acropora lokani Wallace, 1994			•		
Acropora longicyathus (Milne Edwards dan Haime, 1860)	U	•	•		•
Acropora loripes (Brook, 1892)	•	•	•	•	•
Acropora lovelli Veron dan Wallace, 1984	U				
Acropora lutkeni Crossland, 1952	•	•	•	•	•
Acropora microclados (Ehrenberg, 1834)	•	•	•	•	•
Acropora microphthalma (Verril, 1859)	•	•	•	•	•
Acropora millepora (Ehrenberg, 1834)	•	•	•	•	•
Acropora minuta Veron, 2002	•				
Acropora mirabilis (Quelch, 1886)					•
Acropora monticulosa (Brüggemann, 1879)	•	•	•	•	•
Acropora nana (Studer, 1878)	•	•	•	•	•
Acropora nasuta (Dana, 1846)	•	•	•	•	•
Acropora nobilis (Dana, 1846)	•	•	•	•	•
Acropora ocellata (Klunzinger, 1879)		•			
Acropora orbicularis (Brook, 1892)	U				•
Acropora palmerae Wells, 1954	•	•			
Acropora paniculata Verril, 1902	•	•	•	•	•
Acropora papillare Latypov, 1992	•	•	•		
Acropora parahemprichii Veron, 2002	•				
Acropora pectinatus Veron, 2002			•		
Acropora pichoni Wallace, 1999					•
Acropora pinguis Wells, 1950	U				
Acropora plana Nemenzo, 1967			•		•
Acropora plumosa Wallace & Wolstenholme, 1998			•		•
Acropora polystoma (Brook, 1891)	•	•	•	•	•
Acropora pulchra (Brook, 1891)	•	•	•	•	•
Acropora retusa (Dana, 1846)	U				
Acropora robusta (Dana, 1846)	•	•	•	•	•
Acropora russelli Wallace, 1994	•	•			
Acropora samoensis (Brook, 1891)	•		•	•	•
Acropora sarmentosa (Brook, 1892)	•	•	•	•	•
Acropora secale (Studer, 1878)	•	•	•	•	•
Acropora selago (Studer, 1878)	•	•	•	•	•

table continued on next page

Appendix 5.3. *continued.*

Zooxanthellate scleractinia	BALI	KOM	WAK	BNP	DER
Acropora seriata (Ehrenberg, 1834)					•
Acropora simplex Wallace & Wolstenholme, 1998	•		•		•
Acropora solitaryensis Veron dan Wallace, 1984	•	•	•	•	•
Acropora spathulata (Brook, 1891)		•	•		•
Acropora speciosa (Quelch, 1886)	•	•	•	•	•
Acropora spicifera (Dana, 1846)	•	•	•	•	•
Acropora striata (Verrill, 1866)	•		•		•
Acropora subglabra (Brook, 1891)	•	•	•	•	•
Acropora subulata (Dana, 1846)	•	•	•	•	•
Acropora suharsonoi Wallace, 1994	•				
Acropora sukarnoi Wallace, 1997	•				
Acropora tenella (Brook, 1892)			•		•
Acropora tenuis (Dana, 1846)	•	•	•	•	•
Acropora turaki Wallace, 1994			•		•
Acropora tutuilensis Hoffmeister, 1925		•			
Acropora valenciennesi (Milne Edwards dan Haime, 1860)	•	•	•	•	•
Acropora valida (Dana, 1846)	•	•	•	•	•
Acropora vaighani Wells, 1954	•	•	•	•	•
Acropora vermiculata Nemenzo, 1967			•		
Acropora verweyi Veron dan Wallace, 1984	•	•	•		•
Acropora willisae Veron dan Wallace, 1984	•	•			•
Acropora yongei Veron dan Wallace, 1984	•	•	•	•	•
Genus *Isopora* Studer, 1878					
Isopora brueggemanni (Brook, 1893)	•	•	•	•	•
Isopora crateriformis (Gardiner, 1898)				•	•
Isopora cuneata (Dana, 1846)	•	•		•	•
Isopora palifera (Lamarck, 1816)	•	•	•	•	•
Isopora "Komodo"	•	•			
Genus *Astreopora* Blainville, 1830					
Astreopora cucullata Lamberts, 1980	•	•	•	•	•
Astreopora expansa Brüggemann, 1877				•	•
Astreopora gracilis Bernard, 1896	•	•	•	•	•
Astreopora incrustans Bernard, 1896	•	•			•
Astreopora listeri Bernard, 1896	•	•	•	•	•
Astreopora myriophthalma (Lamarck, 1816)	•	•	•	•	•
Astreopora ocellata Bernard, 1896			•		•
Astreopora randalli Lamberts, 1980		•	•	•	•
Astreopora suggesta Wells, 1954	•	•	•	•	•
Family Euphyllidae Veronm 2000					
Genus *Euphyllia* Dana, 1846					
Euphyllia ancora Veron dan Pichon, 1979	•	•	•	•	•
Euphyllia cristata Chevalier, 1971	•	•	•	•	•
Euphyllia divisa Veron dan Pichon, 1980	•	•	•	•	•

table continued on next page

Appendix 5.3. *continued.*

Zooxanthellate scleractinia	BALI	KOM	WAK	BNP	DER
Euphyllia glabrescens (Chamisso dan Eysenhardt, 1821)	•	•	•	•	•
Euphyllia paraancora Veron, 1990	•	•			•
Euphyllia yaeyamaensis (Shirai, 1980)			•	•	•
Euphyllia sp. New	•				
Genus *Catalaphyllia* Wells, 1971					
Catalaphyllia jardinei (Saville-Kent, 1893)		•	•		•
Genus *Nemenzophyllia* Hodgson and Ross, 1981					
Nemenzophyllia turbida Hodgson and Ross, 1981					•
Genus *Plerogyra* Milne Edwards and Haime, 1848					
Plerogyra simplex Rehberg, 1892	•		•	•	•
Plerogyra sinuosa (Dana, 1846)	•	•	•	•	•
Genus *Physogyra* Quelch, 1884					
Physogyra lichtensteini (Milne Edwards and Haime, 1851)	•	•	•	•	•
Family Oculinidae Gray, 1847					
Genus *Galaxea* Oken, 1815					
Galaxea acrhelia Veron, 2002			•	•	•
Galaxea astreata (Lamarck, 1816)	•		•	•	•
Galaxea fascicularis (Linnaeus, 1767)	•	•	•	•	•
Galaxea horrescens (Dana, 1846)		•	•	•	•
Galaxea longisepta Fenner & Veron, 2002	•	•		•	•
Galaxea paucisepta Claereboudt, 1990					•
Family Siderasteridae Vaughan and Wells, 1943					
Genus *Pseudosiderastrea* Yabe and Sugiyama, 1935					
Pseudosiderastrea tayami Yabe and Sugiyama, 1935	•				
Genus *Psammocora* Dana, 1846					
Psammocora contigua (Esper, 1797)	•	•		•	•
*Psammocora decussata*Yabe and Sugiyama, 1937			•		
Psammocora digitata Milne Edwards and Haime, 1851	•	•		•	•
Psammocora explanulata Horst, 1922	•	•	•	•	•
Psammocora haimiana Milne Edwards and Haime, 1851	•	•	•	•	•
Psammocora nierstraszi Horst, 1921	•	•	•	•	•
Psammocora obtusangula (Lamarck, 1816)	•	•	•	•	•
Psammocora profundacella Gardiner, 1898	•	•	•	•	•
Psammocora stellata Verrill, 1868	•				
Psammocora superficialis Gardiner, 1898	•	•	•		•
Genus *Coscinaraea* Milne Edwards and Haime, 1848					
Coscinaraea columna (Dana, 1846)	•	•	•	•	•
Coscinaraea crassa Veron and Pichon, 1980	•				•
Coscinaraea exesa (Dana, 1846)	•		•		•
Coscinaraea monile (Foskål, 1775)	•	•	•		•
Coscinaraea wellsi Veron and Pichon, 1980	•		•	•	•
Genus *Craterastrea* Head 1981					

table continued on next page

Appendix 5.3. *continued.*

Zooxanthellate scleractinia	BALI	KOM	WAK	BNP	DER
Family Agariciidae Gray, 1847					
Genus *Pavona* Lamarck, 1801					
Pavona bipartita Nemenzo, 1980	•	•	•	•	•
Pavona cactus (Forskål, 1775)	•	•	•	•	•
Pavona clavus (Dana, 1846)	•	•	•	•	•
Pavona danai Milne Edwards and Haime, 1860	•		•		
Pavona decussata (Dana, 1846)	•	•	•	•	•
Pavona duerdeni Vaughan, 1907	•	•	•	•	•
Pavona explanulata (Lamarck, 1816)	•	•	•	•	•
Pavona frondifera (Lamarck, 1816)	•	•			•
Pavona maldivensis (Gardiner, 1905)			•	•	•
Pavona minuta Wells, 1954	•	•	•	•	•
Pavona varians Verrill, 1864	•	•	•	•	•
Pavona venosa (Ehrenberg, 1834)	•	•	•	•	•
Genus *Leptoseris* Milne Edwards and Haime, 1849					
Leptoseris explanata Yabe and Sugiyama, 1941	•	•	•	•	•
Leptoseris foliosa Dinesen, 1980	•	•	•	•	•
Leptoseris gardineri Horst, 1921				•	•
Leptoseris hawaiiensis Vaughan, 1907	•	•	•	•	•
Leptoseris incrustans (Quelch, 1886)	•		•	•	•
Leptoseris mycetoseroides Wells, 1954	•	•	•	•	•
Leptoseris papyracea (Dana, 1846)	•	•			•
Leptoseris scabra Vaughan, 1907	•	•	•	•	•
Leptoseris solida (Quelch, 1886)		•	•	•	•
Leptoseris striata Fenner & Veron 2002	U	•	•	•	•
Leptoseris tubulifera Vaughan, 1907					•
Leptoseris yabei (Pillai and Scheer, 1976)		•	•		•
Genus *Coeloseris* Vaughan, 1918					
Coeloseris mayeri Vaughan, 1918	•	•	•	•	•
Genus *Gardineroseris* Scheer and Pillai, 1974					
Gardineroseris planulata Dana, 1846	•	•	•	•	•
Genus *Pachyseris* Milne Edwards and Haime, 1849					
Pachyseris foliosa Veron, 1990		•	•	•	•
Pachyseris gemmae Nemenzo, 1955	•	•	•	•	•
Pachyseris rugosa (Lamarck, 1801)	•	•	•	•	•
Pachyseris speciosa (Dana, 1846)	•	•	•	•	•
Family Fungiidae Dana, 1846					
Genus *Cycloseris* Milne Edwards and Haime, 1849					
Cycloseris colini Veron, 2002		•			•
Cycloseris costulata (Ortmann, 1889)	•	•	•	•	•
Cycloseris curvata (Hoeksema, 1989)	•				
Cycloseris cyclolites Lamarck, 1801	•	•			•
Cycloseris erosa (Döderlein, 1901)	•				

table continued on next page

Appendix 5.3. *continued.*

Zooxanthellate scleractinia	BALI	KOM	WAK	BNP	DER
Cycloseris hexagonalis (Milne Edwards and Haime, 1848)	•				
Cycloseris patelliformis (Boschma, 1923)	•				
Cycloseris sinensis (Milne Edwards and Haime, 1851)	•	•			•
Cycloseris somervillei (Gardiner, 1909)					•
Cycloseris tenuis (Dana, 1846)	•	•	•	•	
Cycloseris vaughani (Boschma, 1923)	•			•	
Genus *Diaseris*					
Diaseris distorta Alcock, 1893	•				•
Diaseris fragilis Alcock, 1893	•				•
Genus *Cantharellus* Hoeksema and Best, 1984					
Cantharellus jebbi Hoeksema, 1993			•		
Genus *Heliofungia* Wells, 1966					
Heliofungia actiniformis Quoy and Gaimard, 1833	•	•	•	•	•
Genus *Fungia* Lamarck, 1801					
Fungia concinna Verrill, 1864	•	•	•	•	•
Fungia corona Döderlein, 1901	•			•	•
Fungia danai Milne Edwards and Haime, 1851	•	•	•	•	•
Fungia fralinae Nemenzo, 1955	•	•	•	•	•
Fungia fungites (Linneaus, 1758)	•	•	•	•	•
Fungia granulosa Klunzinger, 1879	•	•	•	•	•
Fungia gravis Nemenzo, 1955	•	•	•	•	•
Fungia horrida Dana, 1846	•	•	•	•	•
Fungia klunzingeri Döderlein, 1901	•	•	•	•	•
Fungia moluccensis Horst, 1919	•	•	•	•	•
Fungia paumotensis Stutchbury, 1833	•	•	•	•	•
Fungia repanda Dana, 1846	•	•	•	•	•
Fungia scabra Döderlein, 1901					•
Fungia scruposa Klunzinger, 1879	•	•	•	•	•
Fungia scutaria Lamarck, 1801	•	•	•	•	•
Fungia spinifer Claereboudt and Hoeksema, 1987		•		•	•
Fungia taiwanensis Hoeksema and Dai, 1991	•	•			
Genus *Ctenactis* Verrill, 1864					
Ctenactis albitentaculata Hoeksema, 1989	H		•	•	•
Ctenactis crassa (Dana, 1846)	•	•	•	•	•
Ctenactis echinata (Pallas, 1766)	•	•	•	•	•
Genus *Herpolitha* Eschscholtz, 1825					
Herpolitha limax (Houttuyn, 1772)	•	•	•		•
Herpolitha weberi Horst, 1921	•	•	•	•	•
Genus *Polyphyllia* Quoy and Gaimard, 1833					
Polyphyllia novaehiberniae (Lesson, 1831)				•	
Polyphyllia talpina (Lamarck, 1801)	•	•	•	•	•
Genus *Sandalolitha* Quelch, 1884					
Sandalolitha dentata (Quelch, 1886)	•	•	•	•	•

table continued on next page

Appendix 5.3. *continued.*

Zooxanthellate scleractinia	BALI	KOM	WAK	BNP	DER
Sandalolitha robusta Quelch, 1886	•	•	•	•	•
Genus *Halomitra* Dana, 1846					
Halomitra clavator Hoeksema, 1989		•		•	•
Halomitra pileus (Linnaeus, 1758)	•	•	•	•	•
Genus *Zoopilus* Dana, 1864					
Zoopilus echinatus Dana, 1846	•	•	•	•	•
Genus *Lithophyllon* Rehberg, 1892					
Lithophyllon lobata Hoeksema, 1989					•
Lithophyllon mokai Hoeksema, 1989			•	•	•
Lithophyllon undulatum Rehberg, 1892				•	•
Genus *Podabacia* Milne Edwards and Haime, 1849					
Podabacia crustacea (Pallas, 1766)	•	•	•	•	•
Podabacia lankaensis Veron, 2002			•		
Podabacia motuporensis Veron, 1990	•	•		•	•
Family Pectiniidae Vaughan and Wells, 1943					
Genus *Echinophyllia* Klunzinger, 1879					
Echinophyllia aspera (Ellis and Solander, 1788)	•	•	•	•	•
Echinophyllia echinata (Saville-Kent, 1871)	•	•	•	•	•
Echinophyllia echinoporoides Veron and Pichon, 1979	•	•	•	•	•
Echinophyllia orpheensis Veron and Pichon, 1980			•		•
Genus *Echinomorpha* Veron, 2000					
Echinomorpha nishihirai (Veron, 1990)			•		
Genus *Oxypora* Saville-Kent, 1871					
Oxypora crassispinosa Nemenzo, 1979	•	•	•	•	•
Oxypora glabra Nemenzo, 1959	•	•	•	•	•
Oxypora lacera Verrill, 1864	•	•	•	•	•
Genus *Mycedium* Oken, 1815					
Mycedium elephantotus (Pallas, 1766)	•	•	•	•	•
Mycedium mancaoi Nemenzo, 1979	•	•	•	•	•
Mycedium robokaki Moll and Best, 1984	•	•	•	•	•
Mycedium steeni Veron, 2002					•
Genus *Pectinia* Oken, 1815					
Pectinia africanus Veron, 2002	U				
Pectinia alcicornis (Saville-Kent, 1871)	•	•	•	•	•
Pectinia ayleni (Wells, 1935)	•	•	•	•	•
Pectinia elongata Rehberg, 1892			•		•
Pectinia lactuca (Pallas, 1766)	•	•	•	•	•
Pectinia maxima (Moll and Borel Best, 1984)	•	•	•	•	•
Pectinia paeonia (Dana, 1846)	•	•	•	•	•
Pectinia teres Nemenzo and Montecillo, 1981	•		•	•	•
Family Merulinidae Verrill, 1866					
Genus *Hydnophora* Fischer de Waldheim, 1807					
Hydnophora exesa (Pallas, 1766)	•	•	•	•	•

table continued on next page

Appendix 5.3. *continued.*

Zooxanthellate scleractinia	BALI	KOM	WAK	BNP	DER
Hydnophora grandis Gardiner, 1904	•		•	•	•
Hydnophora microconos (Lamarck, 1816)	•	•	•	•	•
Hydnophora pilosa Veron, 1985	•	•	•		•
Hydnophora rigida (Dana, 1846)	•	•	•	•	•
Genus *Paraclavarina* Veron, 1985					
Genus *Merulina* Ehrenberg, 1834					
Merulina ampliata (Ellis and Solander, 1786)	•	•	•	•	•
Merulina scabricula Dana, 1846	•	•	•	•	•
Merulina scheeri Head, 1983		•			
Genus *Boninastrea* Yabe and Sugiyama, 1935					
Genus *Scapophyllia* Milne Edwards and Haime, 1848					
Scapophyllia cylindrica Milne Edwards and Haime, 1848	•		•	•	•
Family Dendrophylliidae Gray, 1847					
Genus *Turbinaria* Oken, 1815					
Turbinaria frondens (Dana, 1846)	•	•	•	•	•
Turbinaria heronensis Wells, 1958		•			
Turbinaria irregularis, Bernard, 1896	•	•	•	•	•
Turbinaria mesenterina (Lamarck, 1816)	•	•	•	•	•
Turbinaria patula (Dana, 1846)	•				•
Turbinaria peltata (Esper, 1794)	•	•	•	•	•
Turbinaria reniformis Bernard, 1896	•	•	•	•	•
Turbinaria stellulata (Lamarck, 1816)	•	•	•	•	•
Genus *Heteropsammia* Milne Edwards and Haime, 1848					
Heteropsammia cochlea (Spengler, 1781)	•	•			
Family Caryophylliidae Gray, 1847					
Genus *Heterocyathus* Milne Edwards and Haime, 1848					
Heterocyathus aequicostatus Milne Edwards & Haime, 1848	•	•			
Family Mussidae Ortmann, 1890					
Genus *Blastomussa* Wells, 1961					
Blastomussa wellsi Wijsmann-Best, 1973					
Genus *Micromussa* Veron, 2000					
Micromussa amakusensis (Veron, 1990)	•	•	•		•
Micromussa minuta (Moll and Borel-Best, 1984)			•	•	•
Genus *Acanthastrea* Milne Edwards and Haime, 1848					
*Acanthastrea bowerbanki*Milne Edwards and Haime, 1851			•		
Acanthastrea brevis Milne Edwards and Haime, 1849	•	•	•		•
Acanthastrea echinata (Dana, 1846)	•	•	•	•	•
Acanthastrea hemprichii (Ehrenberg, 1834)	•	•	•	•	•
Acanthastrea hillae Wells, 1955			•		
Acanthastrea ishigakiensis Veron, 1990	•			•	
Acanthastrea lordhowensis Veron & Pichon, 1982	•			•	•
Acanthastrea regularis Veron, 2002	•		•	•	•
Acanthastrea rotundoflora Chevalier, 1975	•		•		•

table continued on next page

Appendix 5.3. *continued.*

Zooxanthellate scleractinia	BALI	KOM	WAK	BNP	DER
Acanthastrea subechinata Veron, 2002	•	•		•	•
Genus *Lobophyllia* Blainville, 1830					
Lobophyllia corymbosa (Forskål, 1775)	•	•	•	•	•
Lobophyllia dentatus Veron, 2002				•	•
Lobophyllia flabelliformis Veron, 2002	•	•		•	•
Lobophyllia hataii Yabe and Sugiyama, 1936	•	•	•	•	•
Lobophyllia hemprichii (Ehrenberg, 1834)	•	•	•	•	•
Lobophyllia robusta Yabe and Sugiyama, 1936	•	•	•	•	•
Lobophyllia serratus Veron, 2002	U				•
Genus *Symphyllia* Milne Edwards and Haime, 1848					
Symphyllia agaricia Milne Edwards and Haime, 1849	•	•	•	•	•
Symphyllia hassi Pillai and Scheer, 1976			•		•
Symphyllia radians Milne Edwards and Haime, 1849	•	•	•	•	•
Symphyllia recta (Dana, 1846)	•	•	•	•	•
Symphyllia valenciennesii Milne Edwards and Haime, 1849	•	•	•	•	•
Genus *Scolymia* Haime, 1852					
Scolymia australis (Milne Edwards and Haime, 1849)					•
Scolymia vitiensis Brüggemann, 1878			•		•
Genus *Mycetophyllia* Milne Edwards and Haime, 1848					
Genus *Australomussa* Veron, 1985					
Australomussa rowleyensis Veron, 1985	•	•	•		•
Genus *Cynarina* Brüggemann, 1877					
Cynarina lacrymalis (Milne Edwards and Haime, 1848)	•	•	•		•
Family Faviidae Gregory, 1900					
Genus *Caulastrea* Dana, 1846					
Caulastrea curvata Wijsmann-Best, 1972				•	
Caulastrea furcata Dana, 1846	•	•	•	•	•
Caulastrea tumida Matthai, 1928			•	•	•
Genus *Favia* Oken, 1815					
Favia danae Verrill, 1872	•		•	•	•
Favia favus (Forskål, 1775)	•	•	•	•	
Favia helianthoides Wells, 1954					•
Favia laxa (Klunzinger, 1879)				•	
Favia lizardensis Veron, Pichon & Wijsman-Best, 1977	•	•	•	•	•
Favia maritima (Nemenzo, 1971)	•	•		•	•
Favia marshae Veron, 2002	•	•			
Favia matthaii Vaughan, 1918	•	•	•	•	•
Favia maxima Veron, Pichon & Wijsman-Best, 1977	•	•	•		•
Favia pallida (Dana, 1846)	•	•	•	•	•
Favia rosaria Veron, 2002	•				•
Favia rotumana (Gardiner, 1899)	•	•	•	•	•
Favia rotundata Veron, Pichon & Wijsman-Best, 1977	•	•	•	•	•
Favia speciosa Dana, 1846	•	•	•	•	•

table continued on next page

Appendix 5.3. *continued.*

Zooxanthellate scleractinia	BALI	KOM	WAK	BNP	DER
Favia stelligera (Dana, 1846)	•	•	•	•	•
Favia truncatus Veron, 2002	•	•	•	•	•
Favia veroni Moll and Borel-Best, 1984	•	•	•	•	•
Favia vietnamensis Veron, 2002			•	•	•
Genus *Barabattoia* Yabe and Sugiyama, 1941					
Barabattoia amicorum (Milne Edwards and Haime, 1850)			•	•	•
Barabattoia laddi (Wells, 1954)	•		•	•	•
Genus *Favites* Link, 1807					
Favites abdita (Ellis and Solander, 1786)	•	•	•	•	•
Favites acuticollis (Ortmann, 1889)	•				•
Favites chinensis (Verrill, 1866)	•	•	•	•	•
Favites complanata (Ehrenberg, 1834)	•	•	•	•	•
Favites flexuosa (Dana, 1846)	•	•	•	•	•
Favites halicora (Ehrenberg, 1834)	•	•	•	•	•
Favites micropentagona Veron, 2002			•		•
Favites paraflexuosa Veron, 2002	•			•	•
Favites pentagona (Esper, 1794)	•	•	•	•	•
Favites russelli (Wells, 1954)	•	•	•	•	•
Favites spinosa (Klunzinger, 1879)			•		•
Favites stylifera (Yabe and Sugiyama, 1937)	•		•	•	•
Favites vasta (Klunzinger, 1879)		•	•	•	•
Genus *Goniastrea* Milne Edwards and Haime, 1848					
Goniastrea aspera Verrill, 1905	•	•	•	•	•
Goniastrea australensis (Milne Edwards and Haime, 1857)	•	•	•	•	•
Goniastrea columella Crossland, 1948				•	
Goniastrea edwardsi Chevalier, 1971	•	•	•	•	•
Goniastrea favulus (Dana, 1846)	U			•	•
Goniastrea palauensis (Yabe and Sugiyama, 1936)		•			
Goniastrea pectinata (Ehrenberg, 1834)	•	•	•	•	•
Goniastrea retiformis (Lamarck, 1816)	•	•	•	•	•
Genus *Platygyra* Ehrenberg, 1834					
Platygyra acuta Veron, 2002	•	•	•	•	•
Platygyra carnosus Veron, 2002	•				
Platygyra contorta Veron, 1990	•	•	•	•	•
Platygyra daedalea (Ellis and Solander, 1786)	•	•	•	•	•
Platygyra lamellina (Ehrenberg, 1834)	•	•	•	•	•
Platygyra pini Chevalier, 1975	•	•	•	•	•
Platygyra ryukyuensis Yabe and Sugiyama, 1936	•	•	•	•	•
Platygyra sinensis (Milne Edwards and Haime, 1849)	•	•	•	•	•
Platygyra verweyi Wijsman-Best, 1976	•	•	•	•	•
Platygyra yaeyamaensis Eguchi and Shirai, 1977		•	•	•	•
Genus *Australogyra* Veron & Pichon, 1982					
Genus *Oulophyllia* Milne Edwards and Haime, 1848					

table continued on next page

Appendix 5.3. *continued.*

Zooxanthellate scleractinia	BALI	KOM	WAK	BNP	DER
Oulophyllia bennettae (Veron, Pichon & Wijsman-Best, 1977)	•	•	•	•	•
Oulophyllia crispa (Lamarck, 1816)	•	•	•	•	•
Oulophyllia laevis (Nemenzo, 1959)	•	•	•	•	•
Genus *Leptoria* Milne Edwards and Haime, 1848					
Leptoria irregularis Veron, 1990	•				•
Leptoria phrygia (Ellis and Solander, 1786)	•	•	•	•	•
Genus *Montastrea* Blainville, 1830					
Montastrea annuligera (Milne Edwards and Haime, 1849)	•		•	•	•
Montastrea colemani Veron, 2002	•	•	•	•	•
Montastrea curta (Dana, 1846)	•	•	•	•	•
Montastrea magnistellata Chevalier, 1971	•	•	•	•	•
Montastrea salebrosa (Nemenzo, 1959)		•	•	•	•
Montastrea valenciennesi (Milne Edwards and Haime, 1848)	•	•	•	•	•
Genus *Plesiastrea* Milne Edwards and Haime, 1848					
Plesiastrea versipora (Lamarck, 1816)	•	•	•	•	•
Genus *Oulastrea* Milne Edwards and Haime, 1848					
Oulastrea crispata (Lamarck, 1816)	•		•		
Genus *Diploastrea* Matthai, 1914					
Diploastrea heliopora (Lamarck, 1816)	•	•	•	•	•
Genus *Leptastrea* Milne Edwards and Haime, 1848					
Leptastrea aequalis Veron, 2002	•	•			•
Leptastrea bewickensis Veron & Pichon, 1977	•				
Leptastrea inaequalis Klunzinger, 1879		•			
Leptastrea pruinosa Crossland, 1952	•	•	•	•	•
Leptastrea purpurea (Dana, 1846)	•	•	•	•	•
Leptastrea transversa Klunzinger, 1879	•	•	•	•	•
Genus *Cyphastrea* Milne Edwards and Haime, 1848					
Cyphastrea agassizi (Vaughan, 1907)	•		•		•
Cyphastrea chalcidium (Forskål, 1775)	•	•	•	•	•
Cyphastrea decadia Moll and Best, 1984	•			•	•
Cyphastrea japonica Yabe and Sugiyama, 1932	•	•	•		•
Cyphastrea microphthalma (Lamarck, 1816)	•	•	•	•	•
Cyphastrea ocellina (Dana, 1864)	•	•			
Cyphastrea serailia (Forskål, 1775)	•	•	•	•	•
Genus *Echinopora* Lamarck, 1816					
Echinopora ashmorensis Veron, 1990				•	
Echinopora gemmacea Lamarck, 1816	•	•	•	•	•
Echinopora hirsutissima Milne Edwards and Haime, 1849					•
Echinopora horrida Dana, 1846	•	•	•	•	•
Echinopora lamellosa (Esper, 1795)	•	•	•	•	•
Echinopora mammiformis (Nemenzo, 1959)				•	•
Echinopora pacificus Veron, 1990	•		•	•	•
Echinopora taylorae (Veron, 2002)		•	•		•

table continued on next page

Appendix 5.3. *continued.*

Zooxanthellate scleractinia	BALI	KOM	WAK	BNP	DER
Family Trachyphylliidae Verrill, 1901					
Genus *Trachyphyllia* Milne Edwards and Haime, 1848					
Trachyphyllia geoffroyi (Audouin, 1826)	•	•	•	•	•
Family Poritidae Gray, 1842					
Genus *Porites* Link, 1807					
Porites massive		•	•	•	•
Porites annae Crossland, 1952	•	•			•
Porites aranetai Nemenzo, 1955	•				
Porites attenuata Nemenzo 1955		•	•	•	•
*Porites australiensis*Vaughan, 1918	•				
Porites cumulatus Nemenzo, 1955	•		•	•	•
Porites cylindrica Dana, 1846	•	•	•	•	•
Porites deformis Nemenzo, 1955	•				•
Porites densa Vaughan, 1918		•			
Porites evermanni Vaughan, 1907	•	•	•		•
Porites flavus Veron, 2002	•				
Porites horizontalata Hoffmeister, 1925	•			•	•
Porites latistella Quelch, 1886	•	•	•	•	•
Porites lichen Dana, 1846	•	•	•	•	•
Porites lobata Dana, 1846	•		•		
Porites lutea Milne Edwards & Haime, 1851	•				
Porites mayeri Vaughan, 1918					•
Porites monticulosa Dana, 1846	•		•	•	•
Porites murrayensis Vaughan, 1918			•		
Porites napopora Veron, 2002	•				•
Porites negrosensis Veron, 1990	•		•	•	•
Porites nigrescens Dana, 1846	•	•	•	•	•
Porites profundus Rehberg, 1892			•	•	
Porites rugosa Fenner & Veron, 2002	•	•	•	•	•
Porites rus (Forskål, 1775)	•	•	•	•	•
Porites sillimaniana Nemenzo, 1976	•				
Porites solida (Forskål, 1775)	•			•	•
Porites stephensoni Crossland, 1952		•		•	•
Porites tuberculosa Veron, 2002	•	•	•	•	•
Porites vaughani Crossland, 1952	•	•	•	•	•
Genus *Goniopora* Blainville, 1830					
Goniopora albiconus Veron, 2002	•		•	•	•
Goniopora burgosi Nemenzo, 1955	•	•	•	•	•
Goniopora columna Dana, 1846	•	•	•	•	•
Goniopora djiboutiensis Vaughan, 1907	•	•	•		•
Goniopora eclipsensis Veron and Pichon, 1982	•		•		•
Goniopora fruticosa Saville-Kent, 1893	•	•	•	•	•
Goniopora lobata Milne Edwards and Haime, 1860	•	•	•	•	•

table continued on next page

Appendix 5.3. *continued.*

Zooxanthellate scleractinia	BALI	KOM	WAK	BNP	DER
Goniopora minor Crossland, 1952	•	•	•	•	•
Goniopora palmensis Veron and Pichon, 1982	•		•	•	•
Goniopora pandoraensis Veron and Pichon, 1982	•		•	•	•
Goniopora pendulus Veron, 1985	•	•	•	•	•
Goniopora somaliensis Vaughan, 1907	•	•	•	•	•
Goniopora stokesi Milne Edwards and Haime, 1851	•	•	•	•	•
Goniopora stutchburyi Wells, 1955	•		•	•	•
Goniopora tenella (Quelch, 1886)		•	•	•	•
Goniopora tenuidens (Quelch, 1886)	•	•	•	•	•
Genus *Alveopora* Blainville, 1830					
Alveopora allingi Hoffmeister, 1925			•		
Alveopora catalai Wells, 1968		•	•		•
Alveopora daedalea (Forskål, 1775)		•			•
Alveopora excelsa Verrill, 1863	•		•		
Alveopora fenestrata (Lamarck, 1816)	•	•	•	•	•
Alveopora gigas Veron, 1985		•	•		•
Alveopora marionensis Veron & Pichon, 1982	•		•		•
Alveopora minuta Veron, 2002	•				•
Alveopora spongiosa Dana, 1846	•	•	•	•	•
Alveopora tizardi Bassett-Smith, 1890	•	•	•	•	•
Alveopora viridis Quoy and Gaimard, 1833	•				
	406	350	388	370	444

Editor's note 27 August 2012:

The new coral reef species referenced in this chapter has been named '*Euphyllia baliensis sp. nov.*' as described in the newly published paper below:

Turak, E., DeVantier, L. & Erdmann, M. 2012, 'Euphyllia baliensis sp. nov. (Cnidaria: Anthozoa: Sclearctinia: Euphylliidae): a new species of reef coral from Indonesia', *Zootaxa*, no. 3422, pp. 52-61.

Chapter 6

Towards the Bali MPA Network

Putu Liza Mustika and I Made Jaya Ratha

6.1 THE IMPORTANCE OF DATA FOR THE BALI MPA NETWORK

Throughout this report we have presented and explained the results of the Bali Marine Rapid Assessment Program 2011. Based on the status of its coral reefs and reef fishes alone, our conclusion is that Bali warrants a considerable effort with regard to marine conservation. Bali's coral reefs are divided into five major coral community types (Chapter 4) and are generally in good condition (Chapter 3). Bali's reef fish species richness is very high: the second highest in the Asia-Pacific (Chapter 5). On the other hand, the data showed a strong indication of serious overfishing in Bali: we only recorded three reef sharks and three Napoleon wrasses over more than 350 diving hours. All of the above, along with the discovery of 13 new fish species, one new coral species, and 13 coral species suspected to be new to Bali's waters warrants the immediate protection of Bali's marine resources.

Despite its apparent simplicity, the term "Marine Protected Area" (MPA) has several definitions. The International Union for Conservation of Nature (IUCN) defines an MPA as "Any area of intertidal or subtidal terrain, together with its overlying water and associated flora, fauna, historical and cultural features, which has been reserved by law or other effective means to protect part or all of the enclosed environment" (Kelleher 1999). Almost a decade later, the IUCN revised its definition of an MPA to "A clearly defined geographical space, recognized, dedicated and managed, through legal or other effective means, to achieve the long-term conservation of nature with associated ecosystem services and cultural values" (IUCN-WCPA 2008). The Indonesian government has loosely translated the term MPA as an "aquatic conservation area", which is defined as "an aquatic area that is protected and managed with a zoning system to assist in the management of fisheries resources and its environment" (MMAF 2009, Article 1).

Bali as a province currently has one established MPA (the Bali Barat National Park in the Regency of Buleleng) and one declared MPA (the Nusa Penida MPA in the Regency of Klungkung, (see Darma et al.(2010)). Several village level MPAs have been initiated along the coasts of Bali, among others in the Tejakula District. These conservation areas, geographically so close to each other, cannot be managed separately without an understanding of the connectivity between them. To effectively manage such MPAs, the new concept of an "MPA Network" is introduced.

An MPA Network is defined as "A collection of individual MPAs or reserves operating cooperatively and synergistically, at various spatial scales, and with a range of protection levels that are designed to meet objectives that a single reserve cannot achieve" (IUCN-WCPA 2008). An MPA Network should be designed "to restore marine ecosystems and associated populations to their full productivity and diversity" (IUCN-WCPA 2008, p. 24). Eight methods or steps necessary to develop an MPA Network have been identified as follows (UNEP-WCMC 2008):

1. Identify and involve the stakeholders
2. Identify goals and objectives
3. Compile data
4. Establish conservation targets and design principles
5. Review existing protected areas
6. Select new protected areas
7. Implement the network
8. Maintain and monitor the protected area network

In addition, IUCN-WCPA (2008) has defined six best practices for the planning of an MPA network, some of which overlap with the UNEP-WCMC steps:

1. Clearly defined goals and objectives
2. Legal authority and long-term political commitment
3. Incorporate stakeholders
4. Use of best available information and precautionary approach
5. Integrated management framework
6. Adaptive management measures

The Bali MRAP 2011 was a major milestone for the Bali MPA Network. The results covered in this document are a part of the data collection process (#3 in the UNEP-WCMC steps and #4 in the IUCN-WCPA best practices). It is, admittedly, not a complete description of the current condition of coral reefs and reef fishes in Bali; the result of the necessity to collect data rapidly. Rather, the MRAP results serve more as a snapshot of this current condition.

The data on coral reef and reef fishes are important, yet insufficient to understand the overall richness of the coastal and marine biodiversity of Bali. Thus, we have tried to balance this report by incorporating secondary data on several migratory taxa in Section 6.2 below.

6.2 ADDITIONAL INFORMATION ON MARINE MEGAFAUNA IN BALI

6.2.1 Sea turtles

Despite the large number of documents on the green sea turtle trade in Bali (see Adnyana et al. 2010), we found very few publications on the status of sea turtles around the island. Anecdotal information that emerged in the late 20th century led to the belief that Bali had been abandoned by sea turtles, with only a few nesting sites remaining, such as Perancak (Negara Regency) and Pemuteran (Buleleng Regency). However, recent data indicate that Bali still has considerable nesting, and possible feeding, sites in need of protection.

Anecdotal information indicates that Bali seems to be the nesting and foraging site for four sea turtle species: the green turtle (*Chelonia mydas*), hawksbill turtle (*Eretmochelys imbricata*), olive ridley's turtle (*Lepidochelys olivacea*) and leatherback turtle (*Dermochelys coriacea*). Bali has at least 11 sea turtle nesting sites around the island. The KSDA (*Konservasi Sumber Daya Alam* The Agency for Conservation of Natural Resources, http://www.ksda-bali.go.id/?page_id=26) provides a list of managed nesting sites as follows: Kuta (Badung), Lepang (Klungkung), Perancak (Jembrana) and Pemuteran (Buleleng). However, Bali has more unmanaged nesting grounds, scattered across Kedonganan (Jimbaran, Badung), Nusa Dua (Badung), Sanur (Denpasar), Serangan (Denpasar), Saba (Gianyar), Tembok (Karangasem) and Yeh Gangga (Tabanan).

The Bali MRAP 2011 scientists recorded sea turtle sightings at five sites: Terora/Sanur (one large green turtle), Gili Biaha/Tanjung Pasir Putih (Padang Bai, one large hawksbill turtle), Gili Selang (Seraya, one large hawksbill turtle), Menjangan (Anchor Wreck and Coral Garden, two hawksbill turtles). Hawksbill turtles (*Eretmochelys imbricata*) are often found foraging in coral reefs; hence their sighting locations might be foraging grounds for this species. Terora has a considerable sea grass ecosystem; it is thus a candidate as a green turtles' (*Chelonia mydas*) foraging site. In addition, in 2009 a green turtle was released from Serangan Island with a satellite tracker which showed a foraging path around Sanur (UNUD-WWF 2009).

Table 6.1 details the sea turtle species currently found in Bali. This information also indicates the importance of immediate on-site management for sea turtle nesting and foraging grounds, particularly for prominent ones such as Perancak (the Kuta nesting site has been so far managed by Profauna, KSDA and the local coast guard). Consequently, Perancak has been included in the list of important sites to be managed in the Bali MPA Network (see Section 6.3).

6.2.2 Marine mammals

The waters off Bali are apparently suitable for marine mammals, in this case cetaceans (whales and dolphins) and dugongs. The waters off the island support at least 11 species of cetaceans (including two sub-species of spinner dolphins), at least one unidentified species of baleen whale and dugongs (the only member of the Sirenia order in the Indo-Pacific) (Table 6.2).

Marine mammal tourism is increasingly an important economic sector in Bali. The main dolphin watching tourism sites in Bali are in Lovina (Buleleng) and the Peninsula (Badung). The primary target species for both sites are spinner dolphins, although the Hawaiian form (*Stenella longirostris longirostris*) is less common for Lovina (Mustika 2011). No marine mammal hunting occurs in Bali, although we have observed occasional use of stranded whale meat post mortem in several areas.

Lovina is the collective name for several villages westward of Singaraja (Buleleng) which belong to two districts: Banjar and Buleleng. The villages of Temukus and Kaliasem are located in Banjar District. The villages of Kalibukbuk, Anturan, Tukad Mungga and Pemaron are located in Buleleng District. Lovina was the first location for dolphin watching tourism in Bali and Indonesia (Gouyon 2005). The industry started in 1987 when, prompted by foreign backpackers, local fishers in the village of Kaliasem formed a dolphin watching association (Mustika 2011). Since that time, the industry has grown exponentially, now with four dolphin watching associations (Kaliasem, Kalibukbuk, Aneka and Banyualit) and no less than 179 dolphin tour boats available to take tourists every day.

As with many marine wildlife viewing activities (Zwirn et al. 2005; Carlson 2010), dolphin watching tourism in Lovina must be regulated. To date, no formal management regime is in place in Lovina, although Mustika's (2011) research has prompted discussion around sustainable dolphin watching practices among local boatmen. As a result, the boatmen have agreed on three in-principle agreements: 1) turning off the engine (or, if this is impractical, lifting the propeller), 2) keeping their distance from the dolphins and 3) avoiding cutting across dolphins' routes.

The Central Buleleng MPA (which basically covers the whole of Lovina) was declared on 22 August 2011 by the Regent of Buleleng with a special design as a marine tourism park. With this in view, it is imperative to implement best practices within the local dolphin watching industry, although it may take more than one year of continuous community engagement and training before all boatmen put the three agreements (and perhaps other codes of practice to be agreed upon later) into effect.

6.2.3 Sharks and other marine megafauna

Written information on the distribution of other marine megafauna is sparse. However, four species of shark (black tip shark, white tip shark, bamboo shark and Wobbegong shark), four species of rays (*Manta birostris*, eagle ray and blue spotted ray) and the iconic sunfish (*Mola mola*) can be found at Nusa Penida in the Klungkung Regency (Darma et al. 2010). Independent observations by Conservation International and Reef Check Indonesia have moreover confirmed the presence of whale sharks (*Rhincodon typus*) in Nusa Penida. Additional anecdotal information from Yayasan Alam Lestari (LINI Foundation) and Mustika's personal observations also confirmed the presence of this species in Tejakula and Lovina, respectively.

The severe depletion of sharks in Bali's waters is an urgent marine conservation management issue due to the important role sharks play in keeping ocean ecosystems healthy (Stevens et al. 2000; Baum & Worm 2009), and is especially so due to the lost opportunity to establish lucrative shark-based tourism in Bali. While reef shark populations in Bali have already been decimated by fishing, certain fishers in southeast Bali are now heavily targeting deepwater and pelagic sharks, including thresher sharks. In just September and October 2011 alone, up to 4,500 thresher sharks (*Alopias sp.*) were estimated to have been harvested in the waters off Nusa Penida, between Klungkung and Karangasem Regencies (Shingler & Perez 2011). All three species of *Alopias* are currently listed as 'vulnerable' in the IUCN Red List (version 2011.1). Because 90% of the sharks harvested in southeast Bali were pregnant females (Shingler & Perez 2011), it is feared that the current harvest rate will deplete Bali of its sharks in the near future.

Table 6.1. List of sea turtle species and their current nesting and foraging grounds in Bali.

No.	Species	IUCN Red List status	Nesting ground (alphabetical)	Foraging ground (alphabetical)
1	Green turtle (*Chelonia mydas*)	**Endangered (v 3.1)**		Nusa Penida (Klungkung)
				Sanur (Denpasar)
2	Hawksbill turtle (*Eretmochelys imbricata*)	Critically endangered (v 3.1)	Pemuteran (Buleleng) Saba (Gianyar)	Gili Selang (Seraya, Karangasem)
				Menjangan (Buleleng)
				Nusa Penida (Klungkung)
				Padang Bai (Karangasem)
3	Olive ridley's turtle (*Lepidochelys olivacea*)	Vulnerable (v 3.1)	Kedonganan (Jimbaran, Badung)	
			Kuta (Badung)	
			Lepang (Klungkung)	
			Nusa Dua (Badung)	
			Pemuteran (Buleleng)	
			Perancak (Jembrana)	
			Saba (Gianyar)	
			Sanur (Denpasar)	
			Serangan (Denpasar)	
			Yeh Gangga (Tabanan)	
4	Leatherback turtle (*Dermochelys coriacea*)	Critically endangered (v 2.3)	Perancak (Jembrana)	
5	**Unidentified species**		Nusa Penida (Klungkung)	
			Tembok (Karangasem)	

Source: UNUD-WWF (2009), KSDA (KSDA 2009), Darma et al. (2010)

Shark Sanctuary

In the light of recent intelligence received about the island's shark population and the shark fishing around the island, we urge the Bali government to implement legislation to create a shark sanctuary that outlaws the capture or killing of any shark species in Bali provincial waters (inclusion of manta rays in this harvest ban is also strongly encouraged, given the high economic value of manta tourism in Nusa Penida in particular). The creation of a Bali shark sanctuary will be well-received by the international press at a time when Bali is increasingly criticized for its environmental problems, and will prevent even further criticism when information on the thresher shark slaughter is exposed internationally. Moreover, such a move would keep Bali in good stead with its competitor destinations for marine tourism, as many of these (including the Maldives, Palau, Micronesia, the Bahamas, and Guam) have recently declared shark sanctuaries to strong international praise. In October 2011 alone, the Marshall Islands created the world's largest shark sanctuary at 1,990,530 km^2. Bali would be well-served to follow suit,

Table 6.2. List of marine mammal species sighted in Bali since 2001

No.	Species (Latin name)	Species (English name)	IUCN Red list status (v 3.1)	Location	Regency
1a	*Stenella longirostris longirostris*[2]	Hawaiian spinner dolphin	Data deficient	Peninsula	Badung
				Lovina	Buleleng
1b	*Stenella longirostris roseiventris*[2]	Dwarf/Southeast Asian spinner dolphin	Data deficient	Peninsula	Badung
				Lovina	Buleleng
2	*Stenella attenuata*[2]	Pan-tropical spotted dolphin	Least concern	Peninsula	Badung
				Lovina	Buleleng
3	*Grampus griseus*[1,2,4]	Risso's dolphin	Least concern	Peninsula	Badung
				Lovina	Buleleng
4	*Lagenodelphis hosei*[2]	Fraser's dolphin	Least concern	Lovina	Buleleng
5	*Globicephala macrorhynchus*[2,3,4]	Short-finned pilot whale	Data deficient	Lovina	Buleleng
				Serangan	Denpasar
6	*Pseudorca crassidens*[2,5]	False killer whale	Data deficient	Nusa Penida	Klungkung
				Peninsula	Badung
7	*Tursiops sp.*[2,5]	Bottlenose dolphin	Data deficient (*T. aduncus*), least concern (*T. truncatus*)	Lovina	Buleleng
				Nusa Penida	Klungkung
				Peninsula	Badung
8	*Feresa attenuata*[1,2,3,4]	Pygmy killer whale	Data deficient	Peninsula	Badung
				Semawang	Denpasar
9	*Steno bredanensis*[3]	Rough-toothed dolphin	Least concern	Suwung	Badung
10	*Physeter macrocephalus*[1,3,4]	Sperm whale	Vulnerable		Badung, Jembrana, Klungkung
11	*Megaptera novaeangliae*[1,3,4]	Humpback whale	Least concern	Tanah Lot	Tabanan
12	*Balaenoptera sp.*[2]	Baleen whale	Depends on the species	Peninsula	Badung
				Lovina	Buleleng
13	*Dugong dugon*[3,4,5]	Dugong	Vulnerable	Tanjung Benoa	Badung
				Nusa Penida	Klungkung

Note:
[1] found stranded (Mustika et al. 2009)
[2] sighted during boat surveys (Mustika 2011)
[3] Ratha personal data 2011
[4] Marine Mammals Indonesia data
[5] sighted during boat survey (Darma et al. 2010)

noting that a shark sanctuary will not only create a strong positive media impression of the political will to act decisively on serious environmental problems, it will also over time (as shark populations recover) contribute significantly to increasing the value of Bali's marine tourism.

Shark diving tourism has become a popular choice for international divers around the world. A single live shark is worth USD 179,000 for diving tourism in Palau per annum; in very sharp contrast to the worth of a single dead shark which sells for a maximum of USD 274 in shark markets (Vianna et al. 2010). For diving tourism to replace shark fishing, finding the exact location(s) to view sharks (e.g., cleaning stations) is therefore recommended, as tourists would only be willing to pay for expensive shark diving packages if the shark sighting probability is high (see Topelko & Dearden 2005; Vianna et al. 2010). The relative position of the shark viewing station is also important: if viewing stations are too far offshore (so that they are not accessible by day-boats), then shark viewing tourism will not be able to make a significant economic contribution towards the local community (Topelko & Dearden 2005).

Towards the Bali MPA Network

The information contained in this report is considered a sufficient trigger to improve the management of existing MPAs and other conservation regimes. The findings that coral reefs in Bali cluster into five communities (Chapter 4) and that Bali supports the second largest reef fish diversity in Indo-Pacific (after Raja Ampat in West Papua – Chapter 5) warrant the development of an MPA network for the whole island because it will increase its source-sink resilience (UNEP-WCMC 2008). The brief information presented on several marine megafauna, particularly sea turtles and marine mammals (Section 6.2), also reflects the importance of an MPA network in Bali. Migratory species are best served by large MPAs that cover most of the, or the entire, life cycle; this approach is often impractical however, if not impossible. A collection of MPAs that are near to each other and ecologically connected can substitute the role of a large MPA by securing part of species' migratory routes and critical habitats (see Sciara 2007 for cetacean examples).

Admittedly, the report does not cover other important baseline data for an MPA network establishment, such as the distribution of mangroves and basic oceanographic information (particularly mid and bottom water column current patterns), the latter of which is desirable to form a better understanding of the connectivity between MPAs. Socio-cultural and economic analyses are also absent from this document. However, the Precautionary Principle dictates that conservation management should still be installed and implemented despite insufficient data (Lauck et al. 1998).

To complete the milestones towards an MPA network, Conservation International Indonesia has been identifying and engaging with local partners (*inter alia*, local communities, government agencies, NGOs and research institutions). Prior to the 2011 Bali MRAP, in June 2010 Conservation International Indonesia conducted a series of stakeholder meetings to identify MPA priority sites around Bali. The meeting produced 25 priority sites, from which seven MPA candidate sites were later recommended. Table 6.3 presents the nine candidate sites for Bali MPA network, including their ecological characteristics and management status. An additional site (Padang Bai – Candidasa) is added to the list, owing to the area's unique coral reef and reef fish compositions, possibly due to cold-water upwelling which is believed to provide better resilience to climate change. A new coral species *Euphyllia sp.* was found in the waters off Padang Bai – Candidasa (see Chapter 4); this species is currently considered endemic to this region of east Bali. The Bali Barat National Park is also included in the network list for it is an

Table 6.3. Priority sites for MPA network in Bali (clock-wise, eastward)

No.	Site name	Exact location	Biological characteristics	Management status
1	Bali Barat National Park	West Bali, Buleleng	Coral reef, reef fish, sea turtle, cetaceans	An official MPA
2	West Buleleng MPA	Pemuteran, Buleleng	Coral reef, reef fish, sea turtle	Declared as an MPA*
3	Central Buleleng MPA	Lovina, Buleleng	Coral reef, reef fish, cetaceans, whale shark	Declared as an MPA*
4	East Buleleng MPA	Tejakula, Buleleng	Coral reef, reef fish, whale shark	Declared as an MPA*
5	Amed – Tulamben	Karangasem	Coral reef, sea turtle, reef fish, shark	n.a.
6	Padang Bai – Candidasa	Karangasem	Coral reef	n.a.
7	Nusa Penida	Klungkung	Coral reef, mangroves, reef fish, cetaceans, whale shark, sea turtles, shark, manta, sunfish	Declared as an MPA**
8	The Peninsula (including Nusa Dua and Bukit Uluwatu)	Badung	Coral reef, reef fish, cetaceans, sea turtles	n.a.
9	Perancak	Negara	Sea turtles, mangroves	n.a.

Note:
*declared on 22 August 2011
**declared in September 2010

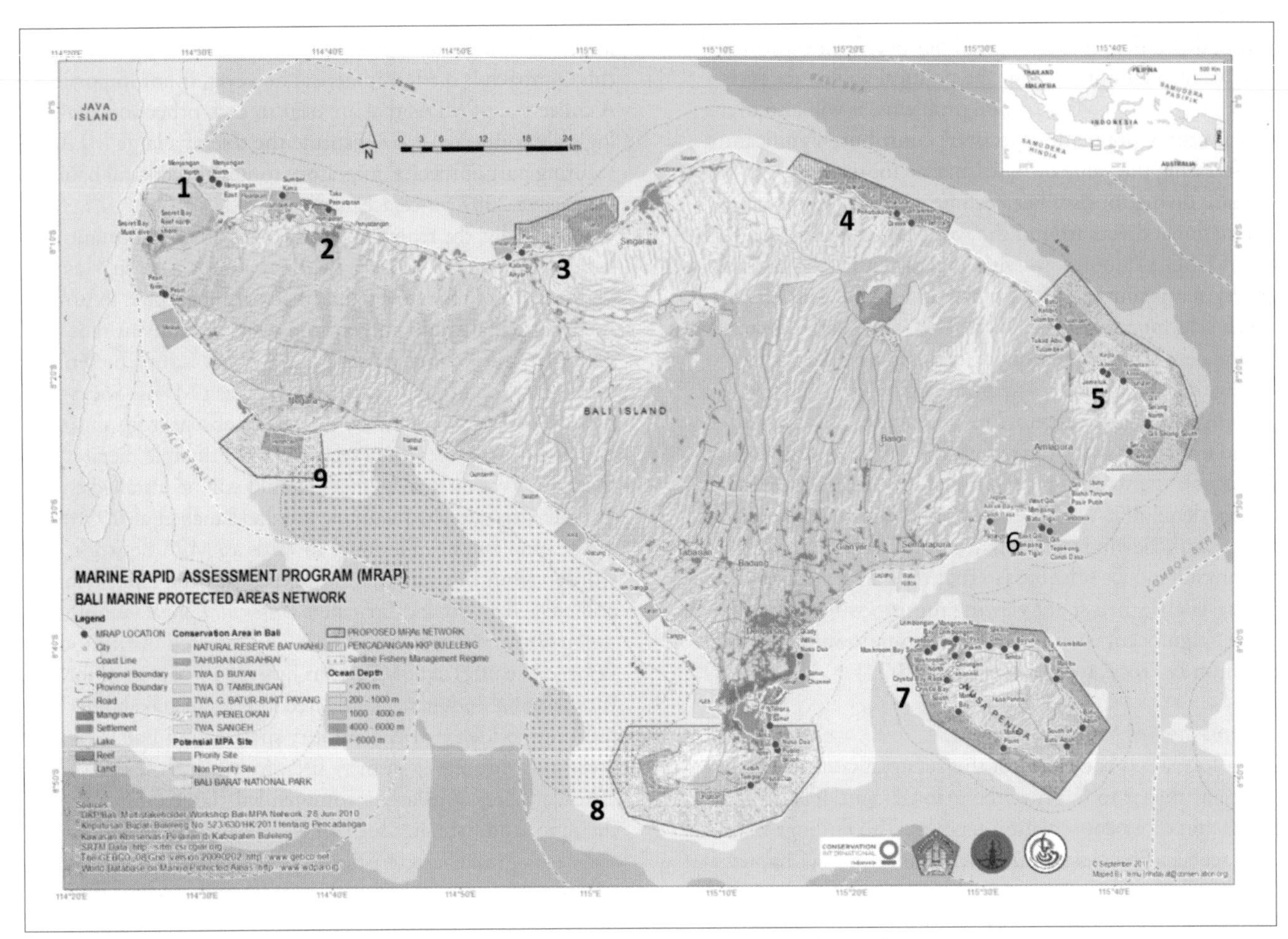

Figure 6.1. Proposed MPAs recommended for inclusion in the Bali MPA Network (see Table 6.3 for MPA names)

important site for coral reefs and reef fishes (Chapters 3–5). In addition to being the first protected area in Bali, the experiences gained from its establishment have much to offer for the development of other MPAs. Figure 6.1 gives a visual distribution of conservation sites listed in Table 6.3.

Of all nine priority sites, only one has an established management regime (Bali Barat National Park). Four other sites (the Buleleng MPAs and Nusa Penida MPA) have been declared as MPAs and now are undergoing planning and zoning processes. The four remaining sites (Amed - Tulamben, Padang Bai-Candidasa, the Peninsula and Perancak) are still devoid of formal management regimes. These sites need to be managed collaboratively by the government, local communities and private sectors, with the help of non-governmental organisations and research institutions. The creation of a provincial level shark sanctuary should also mesh nicely with the island-wide MPA Network concept, as the network will provide additional surveillance and enforcement regimes for the prohibition of shark fishing in Bali waters.

REFERENCES

Adnyana, I. W., Damriyasa, I. M., Trilaksa, I., Ratha, I. M. J. & Hitipeuw, C. 2010, *Laporan Sigi Pemanfaatan dan Perdagangan Penyu di Bali Serta Rekomendasi Pengentasannya (Investigative report on the sea turtle trade in Bali and its alleviation recommendations)*, Faculty of Veterinarian, Udayana University, Denpasar.

Baum, J. K. & Worm, B. 2009, 'Cascading top-down effects of changing ocean predator abundance', *Journal of Animal Ecology*, vol. 78, no. 4, pp. 699–714.

Carlson, C. 2010, *A review of whale watch guidelines and regulations around the world (version 2009)*, International Whaling Commission, Maine.

Darma, N., Basuki, R. & Welly, M. 2010, *Profil Kawasan Konservasi Perairan (KKP) Nusa Penida, Kabupaten Klungkung, Propinsin Bali*, Pemda Klungkung, Kementrian Kelautan dan Perikanan, The Nature Conservancy - Indonesia Marine Program, Denpasar.

Gouyon, A. (ed.) 2005, *The Natural Guide to Bali*, Bumi Kita Foundation, Denpasar.

IUCN-WCPA 2008, *Establishing Resilient Marine Protected Area Networks - Making It Happen*, IUCN-WCPA,

National Oceanic and Atmospheric Administration and The Nature Conservancy, Washington, D.C.

Kelleher, G. (ed.) 1999, *Guidelines for Marine Protected Areas*, IUCN, Cambridge.

KSDA, 2009, *Konservasi Ex-situ (Ex-situ conservation)* [Online], Balai KSDA (Konservasi Sumber Daya Alam) Bali, Available: http://www.ksda-bali.go.id/?page_id=26 [7 September 2011].

Lauck, T., Clark, C. W., Mangel, M. & Munro, G. R. 1998, 'Implementing the precautionary principle in fisheries management through marine reserves', *Ecological Applications*, vol. 8, no. 1, pp. S72–S78.

MMAF 2009, *Decree No. 2/MEN/2009 on the Procedures to Establish an Aquatic Conservation Area*, MMAF, Jakarta

Mustika, P. L. K. 2011, 'Towards Sustainable Dolphin Watching Tourism in Lovina, Bali, Indonesia', Unpublished thesis. James Cook University.

Mustika, P. L. K., Hutasoit, P., Madusari, C. C., Purnomo, F. S., Setiawan, A., Tjandra, K. & Prabowo, W. E. 2009, 'Whale strandings in Indonesia, including the first record of a humpback whale (Megaptera novaeangliae) in the Archipelago', *The Raffles Bulletin of Zoology*, vol. 57, no. 1, pp. 199–206.

Sciara, G. N. d. 2007, *Draft Guidelines for the Establishment and Management of Marine Protected Areas for Cetaceans. UNEP(DEPI)/MED WG.308/8*, United Nations Environment Programme, Palermo.

Shingler, A. & Perez, G. 2011, *Shark Fishing in Nusa Penida September–October 2011*, Denpasar.

Stevens, J. D., Bonfil, R., Dulvy, N. K. & Walker, P. A. 2000, 'The effects of fishing on sharks, rays, and chimaeras (chondrichthyans), and the implications for marine ecosystems', *ICES Journal of Marine Science*, vol. 57, no. 3, pp. 476–494.

Topelko, K. N. & Dearden, P. 2005, 'The Shark Watching Industry and its Potential Contribution to Shark Conservation', *Journal of Ecotourism*, vol. 4, no. 2, pp. 108–128.

UNEP-WCMC 2008, *National and Regional Networks of Marine Protected Areas: A Review of Progress*, UNEP-WCMC, Cambridge.

UNUD-WWF, 2009, *Satellite tracking of DC Bali turtles* [Online], Seaturtle.org, Available: http://www.seaturtle.org/tracking/index.shtml?tag_id=53811&full=1&langn=[7 September 2011].

Vianna, G., Meekan, M., Pannell, D., Marsh, S. & Meeuwig, J. 2010, *Wanted Dead or Alive? The relative value of reef sharks as fishery and an ecotourism asset in Palau*, Australian Institute of Marine Science and University of Western Australia, Perth.

Zwirn, M., Pinsky, M. & Rahr, G. 2005, 'Angling Ecotourism: Issues, Guidelines and Experience from Kamchatka', *Journal of Ecotourism*, vol. 4, no. 1, pp. 16–31.